D0961594

ADVANCE PRAISE FOR

THE REALITY BUBBLE

"In a time of mounting global crisis, the kind of radical curiosity that fills this book—a commitment to probing the unseen, unknowable, and unthinkable—has become essential to our survival. In Ziya Tong's hands, we learn that it can be joyous, too, with thrilling facts, questions, and juxtapositions on every page. A kaleidoscopic guide to everything we're missing."

—Naomi Klein, bestselling author of *No Is Not Enough, The Shock Doctrine*, and *No Logo*

"This book will tear through your preconceptions like a meteor through space. Ziya Tong is a wonderfully erudite companion on a tour of reality, from the very smallest to the size of the universe and everything in between. It's an incredibly illuminating and challenging but ultimately empowering book, and reading it delivers a shock almost on the level as when Neo took the red pill in *The Matrix*. Prepare to have your bubble well and truly burst."

—Rowan Hooper, *New Scientist*

"Humans have a warped perception of reality. Ziya Tong pierces through this collective fog, using a scientific lens to show us our place in the world as it really is. With a journalist's eye for drama, she uses examples from both the history of science and the latest research to expose uncomfortable truths about the shortsighted ways we produce food and energy and dispose of waste, which are jeopardizing life on Earth. Her arguments compel us to look sharp—we remain in this reality bubble at our own peril."

—Kate Wong, *Scientific American*

"Not only is this book a delightful tour of scientific wonders, but it's also a profound meditation on why humans have such a hard time getting out of our bubbles and changing our ways. With grace and humor, Ziya Tong reveals our blind spots—both literal and philosophical—and guides us toward a better future that we can face with both eyes open."

—Annalee Newitz, founder of io9 and author of *Scatter, Adapt, and Remember*

"A book this audacious, inventive, and soulful shouldn't also be so much fun to read. Ziya Tong scours the universe of human knowledge to reframe how we see the world and our place in it."

—Elan Mastai, author of *All Our Wrong Todays*

"*The Reality Bubble* has opened eyes I didn't even know I had. It is so hard to explain how we humans perceive the world, knowing that we can't tell if we all perceive it the same way. This is where Ziya is beyond brilliant: She breaks down how our individual constructions of reality are far more artificial than we realize, which will leave me trying to answer a whole lot of existential questions for some time to come."

—Derek Mead, executive editor, global, VICE

"The world we inhabit is not what it appears to our senses to be. In *The Reality Bubble*, Ziya Tong takes us on a fascinating, whirlwind tour through many unfamiliar aspects of the reality we thought we knew. It is both sobering and mind-blowing to realize how deeply immersed we are in previously hidden realms which science has revealed to us but which most of us ignore at our peril. Tong is an able guide, leading us through the maze of illusions, and helping us to shed our veils of delusion. She illuminates the unseen, and often dangerous, bubbles within which we live out our lives."

—David Grinspoon, Senior Scientist at Planetary Science Institute
and prize-winning author of *Earth in Human Hands*

"Filled with entertaining, often surprising, information, *The Reality Bubble* reveals how science enables us to 'see' beyond the constraints of our physical and psychic barriers and recognize the consequences. Ziya Tong's book should be required reading for all who care about what we are doing to the planet."

—David Suzuki, broadcaster, geneticist, and author of
The Sacred Balance: Rediscovering our Place in Nature

"Ziya Tong takes readers on an important and entertaining scientific journey, as she breaks open all the hidden ways we interact with the natural world every day. Read this book to start seeing—with new eyes—how we can transform our relationship with the extraordinary planet we live on."

—Dr. Jane Goodall, primatologist and conservationist

"Ziya has been reading and thinking about our environments, large and small, for years, and reminds us again and again of the idea of shifting baselines. We accommodate too much of what we experience, moving from surprise to acceptance, recalibrating all the way. This book urges us to be vigilant against that."

—Jay Ingram, television and radio host, and author of *The Science of Why*

"Amid the screaming alarms of the Anthropocene—species collapse, habitat loss, social pathologies—Ziya Tong takes a sledgehammer to the bad ideas that brought us to the brink of apocalypse: unbridled capitalism, technological escapism, species-centrism. Equal parts disaster novel and postmodern cabinet of wonders (and terrors), *The Reality Bubble* makes a passionate, rationalist case for saving the planet before we back off history's cliff, selfie sticks in hand."

—Mark Dery, author of *I Must Not Think Bad Thoughts:
Drive-By Essays on American Dread, American Dreams*

"Human beings are built to detect short-range, near-term threats, and yet our social fabric (and maybe even our species itself) depends on learning to detect the stuff that never directly touches us. In *The Reality Bubble*, Ziya Tong delivers an energetic crash course in this mismatch between our evolutionary gifts and our circumstances. The blind spots she describes are the ones that keep us from internalizing the threat of climate change, the dangers of political manipulation via social networks—even the difficulty of saving for retirement. The physiological and cognitive basis of our reality bubble is something we have to learn about ourselves to get out in front of humanity's biggest problems, and Tong explains the subject, without judgement or derision, in a way that will benefit us all."

—Jacob Ward, NBC News Technology Correspondent,
host of the PBS series "Hacking Your Mind,"
and former editor-in-chief of *Popular Science*

THE REALITY BUBBLE

ZIYA TONG

ALLEN LANE
an imprint of Penguin Canada,
a division of Penguin Random House Canada Limited

Canada • USA • UK • Ireland • Australia • New Zealand • India • South Africa • China

First published 2019

www.penguinrandomhouse.ca

LIBRARY AND ARCHIVES CANADA CATALOGUING IN PUBLICATION

Tong, Ziya, author
The reality bubble / Ziya Tong.

Issued in print and electronic formats.
ISBN 978-0-7352-3556-4 (hardcover).—ISBN 978-0-7352-3557-1 (electronic)

1. Science—Popular works. 2. Science—Miscellanea. I. Title.

Q162.T66 2019 500 C2018-904283-4
C2018-904284-2

Cover and interior design: Lisa Jager
Jacket images: (front) Mike Hill / Getty Images; (back) Xinhui Xu / Unsplash

Printed and bound in Canada

10 9 8 7 6 5 4 3 2 1

For my family.

Up to the Twentieth Century, reality was everything
humans could touch, smell, see, and hear. Since the initial
publication of the chart of the electromagnetic spectrum,
humans have learned that what they can touch, smell,
see, and hear is less than one-millionth of reality.
Ninety-nine percent of all that is going to affect
our tomorrows is being developed by humans using
instruments and working in ranges of reality
that are nonhumanly sensible.

•

—R. BUCKMINSTER FULLER

CONTENTS

INTRODUCTION

IN LIFE, WE ALL HAVE a moment when we wake up to a bigger picture. For Ann Hodges, that moment arrived on November 30, 1954, at precisely 1:46 P.M., while she was lying on the couch. It wasn't so much an epiphany as a painful reality that struck her that day, when a lime-green "cosmic missile" streaked across the clear afternoon sky, crashed through her roof, bounced off a console radio, and flew smack into the side of her body.

As the only known person to be hit by a meteorite, Ann became an instant sensation. By nightfall, hundreds of people, including the national news media, descended upon her backyard, snapping photos of the extraterrestrial object, checking out the damage to her house, and looking in awe and horror at the jet-black, football-sized bruise the impact left on her hip.

Because she was napping, Ann had slept through the fireball's spectacular descent. Witnesses saw it blaze across three states; TVs began scrambling from the alien interference; and the sonic boom jolted a boy right off his bike more than one hundred kilometres away, in Montgomery, Alabama. As for the locals, when the shooting star finally hit ground zero in the town of Sylacauga, most of them thought they'd heard a plane crash or an exploding bomb.

After a few weeks, though, as with all freak accidents, the buzz died down. The reporters packed up and went home and the neighbours returned to their daily lives. And while the meteorite certainly made an impression on everyone that day, only one person's cosmic perspective was forever changed. For Ann Hodges, the universe, with its meteors and comet showers and supernovas, was no longer a separate place somewhere "out there." Oh, no. The cosmos could come right into your house, if it wanted to, and slap you wide awake.

FAR FROM BEING IDYLLIC and tranquil, the heavens are hell. You've got your raging flames, your choking plumes of poisonous gas; darkness, chaos, and violent destruction are pretty much everywhere. In fact, if you look up into the sky tonight in the direction of Sagittarius, just above the archer's arrow, there's a supermassive black hole in our galaxy that is, at this very moment, obliterating everything within its horizon.

That's the universe we live in. But it's not the way it *feels*. That you and I are relatively calm right now, that we aren't in a raw panic at the total and utter mayhem that dangles right above our heads, is because we live in a bubble, a physical one called the atmosphere. From space, this dome is clearly visible. It's a thin, bluish-white film that acts like a planetary force field: it blocks out lethal radiation, maintains temperatures within a tiny range (compared to the extremes of space), and incinerates most of the meteors that would otherwise pulverize the surface of Earth.

As human beings, we all live inside another kind of bubble as well: a psychological one that shapes our ideas about the everyday world. This is our "reality bubble." Just as rocks hurtling at supersonic speed find it hard to penetrate Earth's atmosphere, unwelcome facts and unfamiliar ideas almost never make it through the membrane of the reality bubble. It shields us from thinking about forces "out there" that are seemingly beyond our control and lets us get on with the business of our lives.

But problems arise with inflated certainty, and we see it over and over again. Whether it's real estate bubbles, or stock market bubbles, or political bubbles, being in a bubble means, by definition, that we've got a warped perception of reality. And in the end, all bubbles share the same fate: inevitably, they burst.

So we might do well to remember that even our most stable conceptions of the world can be overturned. For over two centuries, the universe was understood to be governed by Newtonian physics, and then Einstein came along. But it doesn't always take a genius to expand our view of the world. Sometimes it just happens. For Ann Hodges, it happened when a meteorite torpedoed through her ceiling one afternoon. And for you, it may just be the book that you're holding in your hands right now.

HUMANS HAVE A TENDENCY to think we have an accurate picture of the world, but often we're wrong. That's because every person is born with a blind spot. In fact, we have two: one in each eye. In the same way that you would be unable to see all of the movie screen if you were given a crummy seat behind the projector at a theatre, situated at the back of your eyeballs there's an area where light receptors do not grow, because it's the exact spot where the optic nerve jacks into your brain. And yet, despite the fact that the area it eclipses is relatively large (nine full moons in the sky could fit in this broken field of view), most of us never even notice it.

The best way to *see* what you cannot see is with your own eyes. So let's take a look. Cover your left eye and use your right eye to look at the dot above. Now, with your eye still trained on the dot—staying aware of the cross but not focused on it—begin to move your head slowly towards and away from the book. You should notice that at a certain point the cross suddenly vanishes; it disappears from sight. Remarkably, this blank spot doesn't register as some sort of void. Instead, our brain compensates for the emptiness, and with our own perceptual version of Photoshop it even fills in the right background colour. Our blind spots are perfectly camouflaged. We are blind to our blindness.

Now, you might think that a blind spot this obvious would have been detected long ago, but it wasn't until a French physicist named Edme Mariotte was dissecting an eye and came across the bundle of nerves connected to the retina, that he wondered if it might be blocking our sight. Doing some vision tests with his own eyesight, he discovered what was soon to become a mini-sensation in the 1600s. It delighted the nobles of the royal court, who revelled in the magic trick of making each other disappear without blinking an eye. Legend has it that across the Channel, King Charles II would play this visual trick with his prisoners, visually decapitating them with his mind's eye before later executing them in real life.

Of course, blind spots are not only *in* our eyes; they are also in our surroundings. The French for "blind spot" (*angle mort* or death angle) says it all: every year in the United States alone, 840,000 car accidents happen because we can't see something very large driving at us until it crashes into our field of view.

THE PHILOSOPHER LUDWIG WITTGENSTEIN once said that "the aspects of things that are most important for us are hidden because of their simplicity and familiarity." Put another way, we often can't see what's right in front of our noses. We've all experienced it: looking everywhere for your keys when they are staring right at you from the kitchen counter.

Individually, we can be blind to the obvious, but collectively, as a society, we can be blind as well. Here's a curious fact to consider: in the twenty-first century, there are cameras *everywhere*, except where our food comes from, where our energy comes from, and where our waste goes. How is it, then, that the most powerful species on the planet is blind to how it survives?

You might say that modern humans interface with nature as though we live in a bubble. It's the reason why, in the United Kingdom, one in three young adults don't know that eggs come from chickens, a third of children believe that cheese comes from plants, and a whopping 40 percent of youth have no idea that milk comes from cows. For these kids, food comes from where you'd think it comes from: "Duh," the supermarket.

Now, it's not the case that young people aren't smart; it's just that their focus has shifted. The average child in the United States spends forty-five hours a week looking at electronic media and only half an hour of unregulated time outdoors. That being the case, we shouldn't be surprised that the cultural world fogs over the natural one. Immersed in this environment, the average American kid is able to recognize one thousand corporate logos but can't name ten plants or animals native to the area in which they live.

Adults don't fare much better. From inside the bubble, the origin of our greatest source of energy—the fuel that powers our global economy—is also a big unknown. If you take a moment to ask around, you'll soon discover that the average person has no idea what oil is. The liquid we pump into our gas tanks to get to work doesn't come from the pulp of dinosaurs, but every tank of gas *is* powered by a thousand tons of ancient life. So which dead species fuel our daily commute? And what caused those giant graveyards that pressure-cooked into the rich black oil fields we drill for energy?

Finally, we are exceptionally blind to what we waste. From excrement to trash to toxic waste, we live with the illusion that refuse can be made to disappear or, with the push of a button, be magically flushed away. That our waste goes somewhere, that our own pollution finds its way right back into the food we eat, the water we drink, and the air we breathe, is one of the reasons the human race is in such deep shit today.

The kicker is our ignorance as a species would be a lot easier to write off if we weren't also so intelligent. After all, we are the smartest animals on Earth. We are the primates with superpowers. We can fly at the speed of sound and communicate across the planet at the speed of light. Our species has figured out how to hack DNA and change the very codes that govern life.

But the problem is that *life* is disappearing.

Scientists tell us we are currently in the midst of the sixth great extinction. On land, from armadillos to zebras, animal populations are plummeting. In the sea, fish stocks are crashing and coral reefs are bleaching. Glaciers are melting. Droughts are increasing. Wildfires are raging. The population is exploding and the climate is changing. The creep of catastrophe nears day by day, and yet when we reach out our arms . . . it is only to take another selfie.

That somewhere in the back of our minds we know civilization is teetering on the brink explains our cultural obsession with the zombie apocalypse. These dark fantasies don't come from nowhere. We all know that things are going very wrong, but living in a bubble means that, for now, we get to ignore it. Instead, we playfully channel our collective unease, mocking our own fear of a seemingly imminent societal crash. From TV shows to survival guides, we "joke" about building bunkers and stockpiling weapons and food supplies. In cities around the world, tens of thousands gather in "zombie walks" dressed in ghoulish makeup and rags, limping along in a low-rumble chant for one, singular desire.

And what is it that the zombies want? The zombies want *braaaains.*

It's worth asking whether we could fend for ourselves if there were no societal means for survival. Because when you think about it, our system of society works precisely because we conform to it, like brainless zombies. The human population is almost eight billion strong, marching to a capitalist drumbeat of *eat, work, shop, and sleep*. Now, it might be one thing if we loved it, but we don't. I mean, seriously, have you ever met anyone in your life who loves the rat race?

So, given that humanity faces dire consequences *and* that most of us don't even like what we do, the question is: Why do we do it?

The big myth, I will argue, is that we are brought up believing there is no other way. We are simply told that this is how the system works. But what if there is another way? What if this "real world" we're so invested in isn't that real at all? What if we could scrub away the fog of humanity's biggest blind spots so we can see more clearly and begin to uncover what is beyond our reality bubble?

Proust famously said, "The real voyage of discovery consists not in seeking new landscapes but in having new eyes." And so our journey must begin right where we are: by seeing the ordinary, everyday world we live in, in an extraordinary new way.

IN JOHN CARPENTER'S 1988 cult classic sci-fi movie *They Live*, a drifter named John Nada gets hold of a pair of special sunglasses that reveal "truths" that ordinary citizens can't see. Putting them on and looking at magazine ads, billboards, or the TV, he sees their real messages: to obey, consume, conform, and stay asleep.

As a modern parable, the film struck a chord. Its influence can be seen in films, video games, and street art, like Shepard Fairey's Obey series, and in Hal Hefner's political posters and web memes. The film's secret conceit is this: if only a pair of glasses like this existed, people might begin to question why reality is not what it seems.

Luckily, something like that *does* exist.

In this book, we will venture into the unseen world around us, but instead of fictional sunglasses we will be using scientific lenses to bring hidden views to light. That's because scientific instruments are, in a very real way, our new eyes, giving us superhuman abilities to see and hear well beyond what our senses perceive.

On true crime shows, we often catch a glimpse of what modern science can reveal. A nice, tidy living room might appear perfectly ordinary to the naked eye, but once investigators have sprayed luminol—a chemical that reacts with iron in hemoglobin—and flicked off the lights, the chemical's neon-blue glow illuminates blood splatters on the wall, revealing a grisly crime scene.

We have a tendency to think that seeing is believing, but there is so much that we don't see unaided. The same is true for the world around us. Our vision is feeble compared with the most advanced scientific tools. Telescopes allow us to see galaxies over thirteen billion light years away, and using electron microscopes, we can zoom right down to the atomic level to see and touch the very building blocks of our universe.

In the pages ahead then, reality will at times seem bizarre and disorienting. Like falling down a rabbit hole into Wonderland. We will shrink in size, grow into giants, and even find ourselves understanding the languages of other animals. Applying this scientific lens to the world around us radically alters our old ideas about the world, allowing us to question what surrounds us, what sustains us, and, perhaps most importantly, what controls us.

AS A SCIENCE BROADCASTER and journalist, I have spent more than a decade interviewing and learning from the world's top scientists and thinkers. One of the great advantages of working with scientists from many different fascinating fields is that it has given me a broad spectrum of scientific knowledge to draw from, allowing me to share and communicate expertise from a wide range of disciplines. These different disciplines are like pieces in a puzzle. Individually, each gives us a clue as to what's going on, but only by putting them together can we see the bigger picture.

And now more than ever, we need to see clearly, because we are at a critical juncture in human history. Our species is locked on a deadly collision course, one that threatens to extinguish life on Earth precisely because our vision of reality is incompatible with scientific truth. Instead, what we call "common sense" thinking has blinded us for far too long.

In this book, we will examine ten of humanity's biggest blind spots. Section One begins with an introduction to the blind spots we are born with as individuals, and reveals how science and technology allow us to see beyond our biological limits. With this new form of sight, we will journey through the everyday world to uncover what our own eyes are unable to perceive.

In Section Two, we will look at our collective blind spots and investigate how as a society we engage in willful blindness. We'll focus on the most critical aspects of our basic biology—our food, energy, and waste—and see how science has radically transformed the support system our lives depend on, and engineered a world that to the average person is almost entirely opaque.

Finally, in Section Three, we will examine intergenerational blind spots. These are ways of thinking about the world that seem natural or inevitable but are in fact inherited world views passed on from generation to generation. Here we will examine how we navigate the grand dimensions of time and space like the proverbial fish that knows not the water in which it swims.

Carl Sagan once said that "our species needs, and deserves, a citizenry with minds wide awake and a basic understanding of how the world works." This book is a humble effort to respond to that need. So let us begin.

PART ONE

BIOLOGICAL BLIND SPOTS

WHAT SURROUNDS US

1
THE OPEN JAR

Where the telescope ends, the microscope begins.
Which of the two has the grander view?
—VICTOR HUGO

In the blink of an eye, Dondidier was gone, but his disappearance was not a part of the circus act. As the *Hamilton Daily Times* reported on August 16, 1913, detectives and sniffer dogs were quickly dispatched to track down the performer, who vanished two days before opening night. Fortunately, the show was not called off. By Friday evening, the acrobat was spotted by a crew member, hiding inside the main tent. And while the fiasco made the headlines, for the public, the real story was not his mysterious return, it was his worth. The circus star was valued at $500, which, in today's money is more than $12,000; a preposterous amount by all accounts, since Dondidier was just a flea.

A century before the bright lights of Hollywood, the greatest show on earth was tiny: it was the flea circus. The little top was an international sensation, and in cities like New York, Paris, and London, crowds came from afar to watch the parasites perform. There were the ballerina fleas, the sword-fighting fleas, the cannonball fleas, the strongman fleas, the tightrope walkers, the tango dancers, and the trapeze artists. It was here, dazzled by their miniature feats of daring, that audiences applauded the most reviled creature of them all: *Pulex irritans*, the bloodthirsty, plague-carrying, human flea, had catapulted into the spotlight and become a star.

The popularity of the flea circus came, in part, from its well-guarded secret. The big question being: How do you train a flea? Plucked quite literally from the casting couch, the insects were skilled fugitives and could easily hop off the stage and escape. So, when pressed, the flea trainers, or "professors" as they were formally known, revealed a trick for taming the tiny beasts: to keep the animals under control at all times, they held them in an invisible prison.

To do this, the fleas were dropped into a small glass jar and carefully sealed inside. As wingless pests that evolved to leap onto their hosts for a blood meal, fleas have spring-loaded legs that let them jump over one hundred times their own height and the endurance to keep bouncing over thirty thousand times. But inside the jar, their athletic prowess worked against them, because as the fleas shot skyward, they smacked their bodies hard, and repeatedly, up against the lid.

But soon—to avoid the pain—the fleas learned; instead of jumping high, they jumped lower so they no longer ricocheted off the top. At this point, according to the professors, you could leave the lid off forever and the pests would never escape. For the fleas, freedom was only a bounce away, but the trap had been set in their minds.

The story was good. Good enough to fend off the curious—but it also wasn't true. And while flea training may yet hold a lesson for *human* society, it was completely lost on the fleas. That's because behind the scenes, as the "professors" knew full well, the bloodsuckers could not be trained; that if you put a flea inside a jar and remove the lid, a flea, of course, will flee.

But peering through magnifying glasses, eyewitnesses swore that they saw the fleas dancing and juggling at their master's bidding. So the question remains: how did the insects perform the incredible stunts? It turns out, the cheerful spectacle had a dark side. For the fleas, it was torture.

Dressed in pink tutus and glued to tiny parasols, the insects were not willing participants. The gold wire leashes that they wore were harnesses that were used to subject them to noxious conditions. "Soccer-playing" fleas, for instance, played with a tiny cotton ball soaked in citronella, which was repulsive enough to them that they kicked it away on contact. The "jugglers," on the other hand, were held on their backs with glue and the motion of their legs rolled a lint ball above them. As for the musicians in the flea "orchestra," they were tied down to seats on a music box, each with a miniature instrument stuck to its forelegs. Then, with a little tap on the head to each—or sometimes, more sadistically, with a flame lit beneath—they'd begin flailing their free legs about, giving the appearance of waving to the music.

Now before we cue the tiny violin, we should be reminded that to the average person, one flea's life is worthless. Even a hundred lives, or a hundred thousand. We wouldn't blink at a global flea Armageddon; we'd be pleased to be rid of them. But strangely, when people today see "strongman" fleas on YouTube pulling tiny carts, or "acrobat" fleas walking tightropes; when they are onscreen at a scale we can interact with, magnified like micro movie stars, the reaction to these pests changes: *You're hurting the fleas! The leashes are strangling them! This is animal cruelty!*

Keep in mind, in their own homes, chances are these people would crush a flea dead in an instant and fumigate for good measure.

Here's the thing: as giants, human beings have a tendency to treat small life as though it's insignificant. As flea expert and entomologist Tim Cockerill has observed: "Sometimes, in a city like London, you'll see the tiniest speck flying across the room or landing on the table, or in your beer at the pub, and most people don't think of this as a life. They'll just pick it out and flick it away, like it's a bit of dust or soot or whatever, but that's actual animal diversity. If you take a moment to look at that speck, it opens up a whole new world."

And it's true. In fact, whole new species have been discovered in this way.[1]

ROBERT HOOKE WAS AN INTELLECTUAL GIANT, but crippled with scoliosis and Pott's disease, he was also a hunchback. Regarded by some as the Leonardo da Vinci of England, he made a staggering number of contributions in the fields of astronomy, biology, physics, paleontology, and even architecture. Early on, he developed the wave theory of light, proved the existence of air, defined the limits of human vision, discovered and named the cell, deduced that fossils were the remains of once-living things, and proposed the idea, inconceivable at the time, that species could disappear through extinction. But today, he is best known for one iconic drawing: a magnified illustration of a flea.

Folding out over four pages, and "depicted with the anatomical precision of a rhinoceros," as Oxford historian Allan Chapman wrote, the magnified beast was a centrefold from Hooke's 1665 bestseller, *Micrographia*. And while Hooke's notoriously difficult

---

1 Tim Cockerill discovered a new species of parasitic wasp when it "committed suicide" and fell into his cup of tea one day.

personality made him unpopular with fellow academics,[2] his book at least made him very popular with the public. In it, he presented the wonders of the magnified world: illustrations of bee stingers, fly's feet, snail's teeth (they have over twenty thousand of them), and even mites in cheese. The detail of the pictures would still baffle most today, but for people introduced to these "minute bodies" for the very first time, the book was nothing short of mind-blowing.

Because of *Micrographia*, the flea was elevated to a microscopic muse. And inspired by Hooke's illustrations another man set his sights on delving even deeper into the world of the minuscule. Grinding finer and finer lenses until his vision was magnified over 270 times,[3] Anton van Leeuwenhoek was a contemporary of Hooke's whose powerful homemade microscopes were so good they landed him the title of "father" of a new field: microbiology.

With the ability to zoom into the level of a micron, or one-millionth of a metre, Van Leeuwenhoek was able to see well beyond the capacity of the naked eye. And so it was that one day, while examining a few drops of rainwater that had collected in a pot, he made an earth-shattering discovery. Wiggling beneath his eyes, at a stupendously small scale, were little creatures swimming through the liquid. They were smaller than anything he had ever seen. He named them *animalcules*.

---

2 You've likely heard Newton's famous quote: "If I have seen further than others, it is by standing on the shoulders of *giants*." (emphasis mine) It's often cited as a reminder of the power of humility. Except, some scholars today believe it may have been the 17th century version of throwing academic shade. It came from a letter written by Newton to Hooke in which they were embroiled in a scientific feud over credit in the field of optics. And Hooke, it should be noted, was short.

3 Based on his drawings, Van Leeuwenhoek is thought to have made some instruments that could magnify objects up to 500 times.

It's important to keep in mind that what we call microorganisms today did not officially exist in the 1600s. Van Leeuwenhoek was the first to access a world that was previously invisible to the human eye. So when in 1673 he began documenting his findings in a series of letters to the Royal Society in London, leading scientists of the day weren't just skeptical, they thought he was either hallucinating or possibly insane.

What Van Leeuwenhoek had on his side, however, was that he was prolific. And as he began looking closely at everyday things, they transformed into magnified wonders. In 1673, he focused his lens on the life force moving through all of us by putting a drop of his own blood under the microscope. The liquid, it turned out, contained solids: flowing through our veins he saw blood cells, which he described as concave "globules."

In 1677, he spied an entirely new life form and discovered protozoa. Creatures "so small, in my sight, that I judged that even if 100 of these very wee animals lay stretched out one against another, they could not reach to the length of a grain of coarse sand." That same year, he made his greatest personal discovery when he examined another body fluid, his own ejaculate. He became the first person to witness living sperm cells, magnified and "moving like a snake or like an eel swimming in water."

Writing to the Royal Society on September 17, 1683, Van Leeuwenhoek had turned his detective work to dental hygiene. Observing the plaque, or "white matter," between his teeth, he pried opened a portal to a whole new dimension: "I then most always saw, with great wonder, that in the said matter there were many very little living animalcules, very prettily a-moving. The biggest sort . . . had a very strong and swift motion, and shot through the water (or spittle) like a pike does through the water. The second sort . . . oft-times spun round like a top . . . and these were far more in number."

There, in his mouth, he had uncovered a metropolis of life

at the most distant frontier of the microscopic world. They are still the tiniest living beings that we know of today. He had discovered bacteria.[4]

But in the scientific community, there were still strong doubts about Van Leeuwenhoek's brazen claims. In a letter to Robert Hooke, the Dutchman wrote, "I suffer many contradictions and oft-times hear it said that I do but tell fairy tales about the little animals." And so the Royal Society called upon the eminent Hooke to replicate and confirm Van Leeuwenhoek's discoveries.

Hooke had looked through a microscope before, but when he reached Van Leeuwenhoek's magnification, what he saw was baffling and "exceeded belief." And yet it was true. In his letter to the Royal Society, he reported,

> I have here sent the Testimonials of eight credible persons; some of which affirm they have seen 10000, others 30000, others 45000 little living creatures, in a quantity of water as big as a grain of Millet (92 of which go to the making up the bigness of a green Pea, or the quantity of a natural drop of water). . . . If according to some of the included testimonials there might be found in a quantity of water as big as a millet seed, no less than 45000 animalcules. It would follow that in an ordinary drop of this water there would be no less than 4140000 living creatures, which number if doubled will make 8280000 living Creatures seen in the quantity of one drop of water, which quantity I can with truth affirm I have discerned.

Under the microscope's glass lens, a tiny window had swung wide open, and the universe it revealed was gigantic.

---

4 Oral bacteria are prolific: "There are 20 billion bacteria in your mouth and they reproduce every five hours. If you go 24 hours without brushing, those 20 billion become 100 billion!"

WE TEND TO FORGET that on the scale of living things we are massive. To us, reality may appear human-sized, but in truth 95 percent of all animal species are smaller than the human thumb. Even tiny animals like fleas are giants compared to the microscopic life forms that inhabit them. As the old rhyme "Siphonaptera" puts it, "Big fleas have little fleas, / Upon their backs to bite 'em, / And little fleas have lesser fleas, / And so, ad infinitum." In essence, even our pests have pests. Given that, it's worth taking a moment to consider exactly what a "pest" is. The term implies a small creature whose very existence and mode of survival is a nuisance. Fleas are only one of a vast number of species we despise. And for good reason: the rat flea notoriously served as the carrier of the *Yersinia pestis* bacterium that killed millions of people around the world, most notably in connection with the Black Death, the pandemic that peaked in Europe in the fourteenth century.[5] Because of this, some people have questioned if there is even a point to the flea's existence. As one commenter wrote online, "There are those creatures that serve no purpose whatsoever. Fleas are such an example. They don't pollinate any flowers, nor do they prey on any destructive or harmful insects. Instead, they siphon the blood of unsuspecting animals and people all the while passing harmful organisms into their bloodstream!" But the flea is not alone in being deemed "unworthy" of being alive. We hold similar attitudes towards cockroaches, mosquitoes, mites, bedbugs, wasps, ants, silverfish, spiders, flies, and many other unwelcome critters anywhere near our homes. We decide which animals should live and which should die. We divide animals into those

5 "The flea has killed millions around the world . . . and is, indissolubly, connected with the history of Black Death. This disease in man is, in fact, caused—as demonstrated by Yersin and Simond—by the triad: bacterium (Yersinia pestis)/rat/flea (Xenopsylla cheopis)."

we admire or that benefit us—insects that are beautiful or have a "purpose," like butterflies and bees—and those we'd prefer to exterminate, especially where they compete for our food in the realm of agriculture.

As a result, we have launched our own "Black Death," a vicious chemical war against these tiny invaders. Globally, agro-chemicals and pesticides have become a multi-billion-dollar industry that grows year over year.[6] But in our efforts to stamp out unwanted pests, we pour over two million metric tons of pesticides onto our plants and soils every year. Unsurprisingly, we aren't just harming the insects we don't like; we are destroying the insects we do like as well.

Scientists tell us we are witnessing a catastrophic collapse of insect populations. A German study found that on protected nature reserves, insect numbers had plummeted by 80 percent. Rodolfo Dirzo, a Stanford University ecologist, has documented a 45 percent decline worldwide in insect populations over the last four decades. And on the International Union for Conservation of Nature (IUCN) Red List, of the 3,623 invertebrates being tracked, 42 percent are under threat of extinction.[7]

---

6 While pesticide manufacturers have argued that the world will face food shortages without pesticides, scientists have found that the claim is overstated, and that the majority of farms would *increase* productivity if they lowered their use of pesticides.

7 While more scientific research is required, the plummeting numbers are setting off alarms around the world. A recent study in Puerto Rico found that 98 percent of ground insects had disappeared over a period of 35 years. In the canopy, the number was 80 percent. By weight, insects typically outweigh humans seventeen times over. Without them, we can expect catastrophic consequences. That's because insects serve as the foundation of our food chain. If insects decline, a domino effect, known as a "bottom-up trophic cascade," will begin to knock out other species that rely on them.

In our desire to exterminate insects, we've lost sight of how critical they are to human survival, but the ripple effect runs right up the food chain. As British biologist Dave Goulson warns, we "are currently on course for ecological Armageddon. If we lose the insects, then everything is going to collapse." That's because insects not only help with pollination, they are nature's garbage men and recyclers as well. As Goulson notes, "Most of the fruits and vegetables we like to eat, and also things like coffee and chocolate, we wouldn't have without insects. Insects also help to break down leaves, dead trees and dead bodies of animals. They help to recycle nutrients and make them available again. If it weren't for insects, cow pats and dead bodies would build up in the landscape."

We won't be alone in feeling the effects. Already, birds that feed on insects have begun to disappear. The number of birds in Europe plunged by four hundred million in the last three decades. Some migrating songbirds, like the meadow pipit, have seen their populations decline by up to 70 percent.

We do not see it happening, and this is potentially our fatal flaw: we tend not to notice that something is disappearing until it is gone.

IN THE END, what killed the flea circus was the disappearance of its star. The little top reigned glorious for well over a hundred years but was forced to shut its tent flaps when the human flea proved no match, not for insecticides, but for the vacuum cleaner.[8] From a business standpoint, it was the cost of importing fleas that made it impractical. As Professor Tomlin, one of the last great flea trainers noted, "I have offers from all over the world to

---

8 *Pulex irritans* has not gone extinct. It can still be found in Greece, Iran, Madagascar, and even Arizona.

take my show, but you're afraid of one thing, when you get out of the country can you get fleas? I went to Sweden and I had to send to Majorca in Spain to get fleas every fortnight."

We have largely rid ourselves of the human flea, but our bodies continue to host many lesser-known species. Fortunately, both for them and for us, they make their livings quietly as tiny companions that we cannot feel or see. You may want to take a deep breath as you read this, but right now your face is crawling with *Demodex* mites, eight-legged arachnids whose closest relatives are spiders. One study found that by the age of eighteen, 100 percent of people tested are host to the mites.[9] Nestled in the beds of our pores and tucked into our eyelashes, the nocturnal creatures emerge each night, moving at a rate of eight to sixteen millimetres an hour, to feed and search for mates on our faces. Scientists still aren't sure exactly what they eat. It could be the sebum, or oil, our pores secrete, or they could be feasting on meals of dead skin cells or bacteria on our skin. One thing scientists do know: while these mites have mouths, they do not have anuses, and the buildup of food means that when they die they explode a flush of material from their guts, which ends up on our faces. And this fecal matter serves as a home to ever-smaller species, because hitching a ride inside the mites' guts are even more prolific life forms: bacteria.

A little face bacteria is nothing, however, when you consider that humans are *covered* head to toe in microbes. And the diversity of species is absolutely bewildering. Taking swabs from sixty subjects' bellybuttons, researchers at North Carolina State University working on the "Bellybutton Biodiversity Project" found a veritable zoo of bacteria, a total of 2,368 different species, over half of which were previously unknown to science. One person's bellybutton even housed a bacterium only known

---

9 Age appears to be a factor, as babies have fewer mites.

to exist in Japanese soil. He had never set foot in Japan, so how did it get there? Well, bacteria are world travellers. Even in drawing a single breath, as microbiologist Nathan Wolfe has observed, we are sampling a safari of microbial species from around the world: "Dust from deserts in China moves across the Pacific to North America and east to Europe, eventually circling the globe. Such dust clouds harbour bacteria and viruses from the soils where they originated, as well as other microbes they pick up from the smoke of garbage fires or from the mist above the oceans they cross."

Air samples collected by scientists at Lawrence Berkeley National Laboratory found as many as 1,800 bacterial species in the air we breathe. These bacterial life forms are not just on us and around us, they are a part of us. Yale University engineers for instance, have found that a person's "mere presence" in a room adds about thirty-seven million bacteria to the mix, every single hour. What we call our own bodies are in truth only half our own. And while the myth is that bacterial cells outnumber human cells ten to one, recent research has proved that we are a little closer to par. An average human body has thirty trillion human cells and about thirty-nine trillion bacteria cells, meaning we are only slightly outnumbered, by a ratio of 1.3:1.[10]

This, of course, raises the question of who is in charge. Them or us?

In this instance, the human-microbe relationship is not so much parasitic as it is symbiotic. Despite the bad press some germs get, we've learned to live together, for the most part, in relative harmony.[11] At birth, however, we are largely

---

10 Bacterial cells are much smaller than human cells—though there are a lot of them, they make up only about .2 kg of our body weight.

11 There are almost 2 billion species of bacteria, the vast majority of which are harmless to human beings.

bacteria-free[12] and acquire the majority of microscopic hitchhikers along the way in life. This is why, if you take a microbial sample from identical twins, you'll find the microbes that inhabit them have different DNA.

It's becoming apparent that *without* bacteria, our lives would be at risk, because what we call "good microbes," like probiotics, are necessary for a healthy immune system. A species known as *Bacteroides fragilis*, for instance, is found in abundance in the guts of most mammals, including 70 to 80 percent of humans. A molecule on the cell surface, called polysaccharide A, boosts regulatory T-cell production, which in turn prevents inflammation in the gut. Scientists working with mice that were specifically bred to be germ-free found them to have poorly functioning regulatory T-cells, but as soon as *B. fragilis* were introduced to their systems, their health improved and their immunity was restored.

We also call on bacteria to help us perform vital survival tasks like eating. If you're a fan of pasta, pies, or french fries, then pat your belly in thanks to *Bacteroides thetaiotaomicron*. In much the same way that cows have bacteria in their rumens that help them digest the cellulose in grasses, humans rely on *B. thetaiotaomicron* to create the enzymes that let us process starchy plant foods.

But bacteria aren't just in charge of regulating our bodies; they have bigger duties as well. As Rick Stevens, a founder of the Earth Microbiome Project, has observed, "Fifty percent of life on Earth is 'invisible' yet responsible for making the planet habitable." Scientists now know that the smallest life forms on Earth

---

12 Relatively but not entirely. There is bacteria in the placenta. "Scientists have spotted bacteria in amniotic fluid, blood in the umbilical cord, the membrane that surrounds the fetus and even babies' first poop."

are responsible for engineering planetary-scale systems, including the very air we breathe and the food we eat. And while humans walk around like we're the most powerful creatures on the planet, in reality it is the microbes that are running the show.

For starters, they produce the gas that is vital for multicellular life—oxygen. And while we are taught that oxygen is exhaled primarily by trees, in fact, only 28 percent of the gas is exhaled from rainforests. The vast majority of oxygen is created in the ocean, by phytoplankton and algae. The source of this photosynthesis is one and the same, however, as both land plants and algae have something in common: they were once hijacked by bacteria.

More than two billion years ago, cyanobacteria evolved an extraordinary superpower: the ability to turn sunlight into food. Using the energy from our nearest star, they began converting water and carbon dioxide into sugars, splitting the remaining oxygen off as by-product. Over time, some species of these cyanobacteria remained aquatic and stayed free-living and independent in the ocean,[13] while others were absorbed by algae and became permanent residents housed inside their organelles, known as chloroplasts.[14] As algal species evolved and migrated onto land, they became the ancestors of modern trees and plants. Which means that these tiny and very ancient engineers sit at the controls of all photosynthesizing plants. And it is they who are responsible for *all* of the oxygen we breathe.

At our feet lies another wildly overlooked ecosystem. Soil is home to a third of all life on the planet, and it is buzzing with

---

13 Today, one particular species does this brilliantly. Described as "the most important microbe you've never heard of," *Prochlorococcas* is responsible for manufacturing a full 20 percent of the oxygen we breathe.

14 Like mitochondria, chloroplasts have their own DNA that comes from the cyanobacteria.

biodiversity. Just a single teaspoon of garden soil contains a population of about a billion bacteria. In terms of biomass, that's the equivalent of about two cows per acre. One handful of forest soil contains more microbes than there are people on Earth, and one kilogram of healthy soil contains more microbes than all the stars in our galaxy. Van Leeuwenhoek could never have dreamed how vast the universe under the microscope would prove to be. But even today, more than three centuries after Van Leeuwenhoek's first discovery, much of this subterranean cosmos of bacteria, archaea, fungi, protozoa, algae, and viruses remains unexplored. So far, only 0.001 percent of microbial species are known to science.

Soil, of course, is critical for food. Without good soil we'd starve. And today, we understand one of the key roles certain bacteria play with respect to plant growth. That's because plants, like all living beings, need nitrogen for their DNA. In the soil, these bacteria have the ability to take atmospheric nitrogen, which is a gas, and "fix" it so that it turns into a form, like ammonia, that plants can use. In essence, nitrogen-fixing bacteria are like tiny "soluble bags of fertilizer" in the soil, feeding the plants their chemical nutrition and in turn enriching every animal on the food chain.

Beyond their habitats on land and in the oceans, bacteria have also been found swirling high up in the atmosphere. Travelling with NASA's hurricane researchers, scientists sampled a cubic metre of air at 33,000 feet and netted over 5,100 species. Our planet is surrounded by a literal bubble of bacteria. Right now, we are only just beginning to find out what these tiny beings are doing up there. Some scientists believe they play an active role in creating clouds and seeding rainfall, while others say they may be recycling nutrients high up in the atmosphere. There is one thing at least we know for certain: far from being insignificant, the smallest life forms on Earth play a critical role in engineering the planet's life-support systems. We have long been blind to the invisible services that bacteria provide, but in truth, we owe them our lives.

OUR FIRST BLIND SPOT is that reality is not human-sized. What we call reality is only a tiny sliver in the grand scheme of things. And while we seldom think about size, size is arguably the most important attribute of an animal's existence: it shapes where, how, and even for how long[15] we live on this planet. When it comes to life on Earth however, size does have its limits.

The parasitic wasp called a fairyfly, for instance, is just two hundred microns across. That's about the size of an amoeba, meaning a family of five of these tiny wasps could fit comfortably on the period at the end of this sentence. But what's incredible about the fairyfly is that, unlike an amoeba, it is not a single-celled organism. It's a complex multicellular life form that has managed to squeeze an incredible amount of biological material into an unbelievably minute package. Inside their bodies, these animals have the basic biological architecture of a beating heart, wings, legs, a digestive system, and a functioning brain. So how does it all fit? For the fairyfly, being small comes at a hefty price, and they pay it in brain cells.

Scientists have discovered that by the time they're adults, fairyflies have sacrificed the nuclei in 95 percent of their neurons, which is where the genetic material is stored in the cell. What that means is, for insects, going even smaller becomes next to impossible. For brainless bacteria, there is still space to shrink. While only five fairyflies could fit on a period at the end of a sentence, hundreds of thousands of single-celled bacteria could occupy the same space. When it comes to size, then, bacteria guard this final frontier. Multicellular life cannot get smaller because there's not enough

15 Small animals tend to live shorter lives.

room for its essential ingredients: proteins and DNA. Meaning life, quite literally, cannot squeeze itself in.

On the opposite end of the spectrum there are the giants: the multicellular animals that operate at our size and the few that are even bigger. So, what then are the limits of large living things? Why are there no real-life King Kongs,[16] Godzillas, or fifty-foot women? The first person to tackle that question was, fittingly, a kind of Goliath himself: the famed stargazer and scientific revolutionary Galileo Galilei.

What Galileo realized was that size not only matters, it can be a matter of life or death. In *Discourses and Mathematical Demonstrations Relating to Two New Sciences*, he wrote, "Who does not know that a horse falling from a height of three or four cubits will break his bones, while a dog falling from the same height or a cat from a height of eight or ten cubits will suffer no injury? Equally harmless would be the fall of a grasshopper from a tower or the fall of an ant from the distance of the moon." In essence: Why would a big animal fall to its death while a small animal could walk away without injury?

Galileo's brilliance was in realizing that if you continued scaling an animal up, at a certain point it would begin to break under its own weight. Just as a tree would no longer be able to support the heft of its massive branches, a fifty-foot giant could not take a step without cracking the bones in her limbs.[17] For the behemoths on Earth, then, it's the laws of physics, and gravity in particular, that puts a limit on things.

---

16 The largest recorded ape primate in the fossil record was *Gigantopithecus blacki*, a three-metre- tall ape. It was doomed by its size in a different manner, however. During the Ice Age, the food supply became insufficient to support the giant ape.

17 For more on size I direct the reader to J.B.S. Haldane's paper "On Being the Right Size."

The observant among you however, must be thinking, what about dinosaurs, or whales? The biggest sauropods were as tall as a five-storey building, and even blue whales measure about the same as three school buses parked end to end. So how come they are so big? It turns out, these massive animals evolved some impressive workarounds.

Dinosaurs got around the heavy bone problem by becoming pneumatic. The titanic reptiles, like their bird descendants today, had light, hollow bones with large air pockets inside them. In fact, 10 percent of *T. rex*'s body volume was air, and in studying sauropod skeletons, scientists have discovered that their bones were up to 90 percent air by volume. Whales solved the problem by evolving in water. Like all living things, a whale's cells contain saline. In the simplest terms, being primarily *made of salty water* and swimming in salty water allows these leviathans to grow to massive sizes and weigh up to 144 metric tons, because, living in the ocean, they're essentially weightless.[18]

There is, however, another invisible medium that can affect an animal's size, and, like water for whales, it's something we barely notice: the air. That vaporous cocktail we all breathe has changed significantly over the ages. And, along with it, so has the size of life.

If you could hop in a time machine and turn the dial back to between one hundred and four hundred million years ago, like Alice, you would emerge into a gargantuan Wonderland. Because this was the age of giants. In this ancient world, mushrooms rose to the height of houses, hawk-sized dragonflies swooped through

---

18 Another factor that scientists believe may affect how marine animals grow to be big in water has to do with loss of heat. Marine mammals grow bigger and have more blubber, as they increase their volume to surface area. This allows them to generate more heat but lose less of it through the surface area of their skin.

the skies, and even dinosaur fleas were ten times larger than their modern counterparts.

Invertebrates were free to grow because for them the weight of bones was not an issue. But there was something else that limited their growth. Kirkpatrick Sale, the author of *Human Scale*, describes the problem like this: "If an earthworm were ten times bigger, its weight would be a thousand times greater, and its need for air a thousand times greater, but the surface area through which it absorbs oxygen would only be a hundred times greater, so it would get only a tenth of the air it needed and would immediately die." So how did prehistoric worms grow in size and still manage to survive? The answer was the concentration of oxygen. In our atmosphere today, oxygen makes up 21 percent of the air, but during the Carboniferous era[19] its concentration was much higher, at 35 percent. For animals like worms that breathe not through their mouths but through pores in their skin, each breath packed a more powerful punch and delivered enough oxygen for them to survive.[20]

We can count ourselves lucky today that there aren't dog-sized cockroaches scuttling through our kitchens. That's because insect gigantism came to an end when another animal rose to prominence. One hundred and fifty million years ago, dinosaurs evolved into a new kind of flying predator: birds. For insects trying to make a quick getaway, the slight and streamlined among them fared better than the big and bulky. Evolution

---

19 The Carboniferous era was specifically from the Devonian period 358.9 million years ago, to the beginning of the Permian period, 298.9 million years ago.

20 Having a larger volume to surface area also meant that the oxygen amount would still be relative to body size, so the animals wouldn't die from oxygen toxicity either.

favoured a smaller body size for escape, and insects began to shrink.[21]

The size of a species is not accidental. It's a fine-tuned interaction—a back and forth—between a species and the world it inhabits. Over large periods of time, size fluctuations, from dwarfism to gigantism, have often signalled significant changes in the environment. Generally speaking, however, over the last five hundred million years, the trend has been towards animals getting larger. It's particularly notable in marine animals, whose mean body size has increased 150-fold in this time.[22]

But we are beginning to see big changes again. Scientists have discovered that many animals are shrinking.[23] Around the world, species in every category—fish, bird, amphibian, reptile, and mammal—have been found to be getting smaller, and one key culprit appears to be the heat.[24] Animals living in the Italian

---

21 There are many interesting incidences of size changes due to environment. Of note, there is Foster's rule, which states that in island environments large animals tend to develop smaller bodies due to restricted food sources, and small animals tend to get larger due to limits in predation. An example of this can be found in mammoths. A mammoth species that lived in Crete, 3.5 million years ago, stood about only one metre at the shoulder.

22 Looking at more than 17, 000 marine species, researchers have found that since the time the animals first evolved, body volumes have increased by five orders of magnitude.

23 As always, there are exceptions. For instance, climate change is making wolf spiders bigger.

24 Historically, scientists have documented mammalian dwarfing during warming periods in the Earth's history. And during the Palaeocene-Eocene Thermal Maximum, a warming phase of three degrees that took place fifty-five million years ago, some mammals shrank by up to a third, while insects like beetles, ants, and bees shrank by three-quarters.

Alps, for example, have seen temperatures rise by three to four degrees Celsius since the 1980s. There, even at an altitude of one thousand metres, heat waves have spiked the alpine temperatures to as high as 30°C. To avoid overheating, chamois goats now spend more of their days resting rather than foraging, and as a result, in just a few decades, the new generations of chamois are 25 percent smaller, and are dwarfs by comparison. Underwater too, sea temperatures have begun to rise, one consequence of which is that the water holds less oxygen and becomes more anoxic. Scientists studying six hundred species of fish say that big size changes are coming and that by 2050 fish will have shrunk by as much as a quarter.

Shrinking potentially signals an even bigger problem: a population crash. Looking at commercial whaling data over four decades, researchers documented that sperm whales shrank substantially—by four to five metres—in the years before their populations collapsed. For biologists, then, shrinking is like an early warning system, alerting us that a species may be in trouble.

But not all animals are shrinking. Domestic species that we raise for food, like pigs and cows, for instance, are growing faster and larger than at any time in history. Since the 1930s, turkeys have more than doubled in size, and since the 1950s, broiler chickens have quadrupled.

To track the changes, Canadian researchers have continued raising unmodified chicken lineages and have measured them against our modern Frankensteins. Like living, breathing, chicken time capsules, these "benchmark strains" are still being bred. This allows researchers to measure commercially selected breeds, like the 2005 Ross 308 Broiler, against older genetic strains. Fed the same food, and measured at the same age, the 1957 strain weighed in at 905 grams, the 1978 strain weighed 1,808 grams, while the 2005 strain weighed 4,202 grams. The difference is enormous. Compared to birds from the 1950s, today's modern broilers have breasts that are 80 percent larger and have increased overall in size by 400 percent.

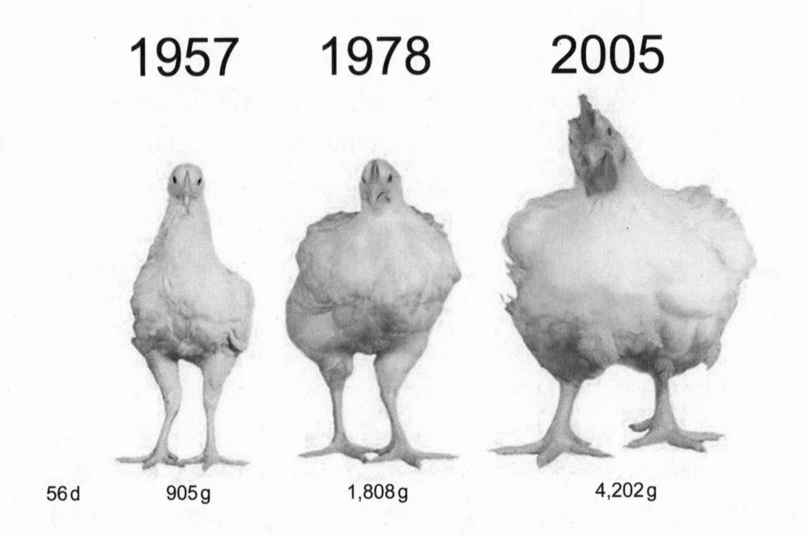

There is a consequence to this. As we've deliberately grown larger animals for food, our appetites have grown as well. In 1960, the average American ate 12.7 kilograms of chicken a year, today that number has jumped to 40.8 kilograms, more than three times as much.[25] Unsurprisingly, as the beneficiaries of all this cheap meat, humans have also begun to change in size. Over the last 150 years, which is, relatively speaking, a short period, human height has increased dramatically. In industrialized countries, where is food abundant, we've grown taller by ten centimetres. Not only have we expanded upward though, we've expanded outward as well, and every country on Earth

---

25 The average consumption of both red meat and poultry in 1960 was 75.3 kilograms, in 2017 it was projected at 98.8.

has seen its obesity rates rise.[26] In total, 2.2 billion people worldwide are classified as overweight or obese, and adults are three times more likely to be obese than they were back in 1975. Today, wild animals world over are shrinking, but human beings and our domesticated animals are ballooning in size.

GALILEO WAS THE FIRST person on Earth to glimpse the colossal scale of reality.[27] Known today as the Father of Science, he was not only the first person to burst open the heavens with a telescope, he was also the first man to peer into a microscope and document the humble flea. It was Galileo's good fortune to be alive at a time when glass-making was flourishing, in particular the craft of making spectacles. Then, as now, people in their forties often developed presbyopia, a condition where the lens of the eye loses flexibility with age, making it more difficult to read. In nearby Holland, the Dutch had become masters at grinding lenses to make reading glasses, and it was these spectacle makers who crafted the first rudimentary instruments that allowed us to bring into focus scales that were previously unseen.

Their intent may have been to remedy poor vision, but the spectacle makers inadvertently did much more. By boosting our vision, they revealed that humanity had been oblivious to two vast scales that secretly co-existed alongside our own. The macro and micro worlds were now made visible, and with this new and improved sight came the realization that we inhabit not only one reality, but three.

---

26 Twenty percent of Pacific Islanders living into Tonga and Tuvalu are classified as obese, and even North Korea has seen a gain of 1 percent.

27 The first published observations using a microscope were in Galileo's *Apiarium* in 1625. He first observed the flea with a microscope in 1624.

For the first time in history we could extend our human senses. And because of that, the first microscopes and telescopes were considered almost magical inventions. Spectacle-making was a secretive and competitive trade, and patent claims to these first inventions are still contested. The design of the first simple and compound microscopes is, however, generally attributed to the spectacle maker Zacharias Janssen, who began developing his new tools in 1590, and the first patent for the "spyglass," or telescope, was filed eighteen years later, in 1608, by master lens grinder and spectacle maker Hans Lippershey.[28]

Galileo was a scientist and not a spectacle maker, but once he wrapped his genius around how the scopes were made, he quickly improved upon both designs. In 1609, he created a device that he named the *occhiolino*, or "little eye," a microscope that could magnify up to thirty times, ten times more than Janssen's design. And that same year, he built his first telescope, a three-powered spyglass that rivalled Lippershey's invention. By August 1609, he surpassed even this with a new prototype telescope, an instrument that magnified eight times, which he presented to the Venetian senate. And by October or November, he had constructed a twenty-powered telescope, and it was this one that he trained upon the skies.

Human vision may be limited, but it's incredible when you consider what the naked eye can see even without the aid of equipment. On a clear night, a person with good eyesight can detect the flicker of a single candle flame[29] 2.76 kilometres away.

---

28 There are three people associated with the invention of the telescope as two patents were filed within weeks of each other. Zacharias Janssen is often also cited as an inventor. Early telescopes were very simple, they were made of two pieces of glass held apart to magnify distant objects.

29 Our ability to detect light is so powerful, that scientists recently discovered that, up close, we can even detect the faintest glimmer from the light of a single photon.

But, depending on the size of an object, or its brightness, we can actually see much farther away than that. The moon, for example, is 385,000 kilometres away, and our own sun is so bright that even at a distance of 150 million kilometres away it can blind us. As for the farthest single object we can see without a telescope? It is Saturn, which is 1.5 billion kilometres away. We can even see a galaxy outside of our own: Andromeda, which shines with the light of a trillion stars. There, in the distance, it flickers like a candle at 2.5 million light years, or twenty-five quintillion kilometres, away.

And all of this comes standard with our basic vision, which we test by looking at a pyramid of black letters known as the Snellen chart. Good visual acuity is the ability to accurately make out the tiny letters on the chart's eighth row, or what we call 20/20 vision. Even in ancient times, sharp eyesight was highly valued. It goes without saying that in selecting the best warriors and hunters, it was crucial to weed out those who couldn't spot the enemy or prey. But our ancestors had a different kind of eye test, however, one that took place not in an optician's office but outside at night, under the canopy of stars.

The asterism called the Big Dipper hangs in the constellation of Ursa Major. It consists of seven points of starlight and forms what looks to us like a giant ladle in the sky. Zooming in to the second star from the left on the handle, you will spot Mizar, twinkling in from seventy-eight light years away. But doubled with Mizar is a dimmer star that's three light years behind it. We call it Alcor, but to Sufi astronomers it was known as Al-Suha, or "the forgotten one." For the ancient Persian army—and some say, on the other side of the world, Native Americans—Alcor was nature's Snellen chart, and the ability to distinguish between the optical double stars was the test of perfect vision.[30]

---

30 The ancient Mizar test has been found to be the modern tested equivalent of 20/20 acuity.

With good vision so highly prized by armies, it was no surprise that Lippershey's spyglass was an instant hit with the Dutch army. Galileo too had entrepreneurial designs for his telescope, which he pitched to the Venetians. "The power of my *cannocchiale* [telescope] to show distant objects as clearly as if they were near should give us an inestimable advantage in any military action on land or sea," he assured the Doge. "At sea, we shall be able to spot their flags two hours before they can see us; and when we have established the number and type of the enemy craft, we shall be able to decide whether to pursue and engage him in battle, or take flight. Similarly, on land it should be possible from elevated positions to observe the enemy camps and their fortifications."

In the end, it wasn't Galileo's ideas for military strategy but something serendipitous that occurred one evening while he was sitting outside and relaxing that forever changed how we see the universe. Instead of training his telescope on the spires of the city, Galileo arced it upward and pointed it into the sky. Through the lens, he began examining the biggest and most luminous object in the night sky, the moon. And what he saw was not at all what he expected. The moon, that perfect sphere of the heavens, was not just a smooth, glowing orb. Looking closer now he could see that it had craters. It had mountains. It had valleys and landforms that were similar to what we had on Earth. The moon, he was shocked to discover, had a landscape. And for Galileo, this was a total revelation.

Pointing his telescope up each night in "infinite amazement," he soon began focusing on other celestial bodies. It was with his observations of Venus that our understanding of our place in the universe changed. What he noticed was that Venus had a shadow, and that, similar to the moon's phases, it changed from a crescent back to a full, shining disc when it faced the sun. For Galileo, this could mean only one thing. Venus was not just a "wandering star"[31]; it had a path. Moreover, this path was not orbiting Earth. It was orbiting the sun.

It was, in every sense of the word, a revolutionary discovery. Until that time, we had believed that the universe revolved around us. Galileo's evidence shattered that idea and proved the Copernican theory of heliocentrism,[32] which placed the sun rather than Earth at the centre of the universe. But there would be no fanfare for Galileo's discovery. For the Church, the observation was dangerous. In the Bible, humanity had clearly been placed at the centre of the universe by God. To believe Galileo implied that the words in the Holy Scripture were false.

So in 1616, the astronomer was summoned by the Roman Inquisition and investigated for heresy. Nicolaus Copernicus's book *On the Revolutions of the Celestial Spheres* had already been banned, and it was decreed that Galileo too was to be silenced. He could no longer make any suggestion, in speech or in writing, that Earth moved around the sun. It was a remarkable moment, because, while we've always thought that seeing is believing, the Church was insisting that we *disbelieve* what we could see with our own eyes. Galileo had discovered a blind spot, but the Church wanted people to stay blind. In the short term, Galileo acquiesced, but sixteen years later he would stand trial again.

---

31 Planets were not differentiated from stars, other than the fact that they seemed to "wander."

32 The seven axioms of heliocentrism are: "(1) There is no one center in the universe; (2) The Earth's center is not the center of the universe; (3) The center of the universe is near the sun; (4) The distance from the Earth to the sun is imperceptible compared with the distance to the stars; (5) The rotation of the Earth accounts for the apparent daily rotation of the stars; (6) The apparent annual cycle of movements of the sun is caused by the Earth revolving round it. (7) The apparent retrograde motion of the planets is caused by the motion of the Earth from which one observes."

Since the days of the first telescopes, our scientific sight has sharpened tremendously, and today we can see so far into the distance we are looking back in time at the universe's beginnings. Across the globe, hundreds of observatories dot the planet, peering into the night like big white robotic eyes. We've built them in cities, on mountaintops, in remote deserts, and we've even sent telescopes out into space. What this extraordinary level of vision means is that we only need to pick a point in the sky and then simply wait.

In September 2003, NASA astronomers did just that. They trained the Hubble Ultra Deep Field telescope on a patch next to the moon that appeared completely empty, not a star visible to the naked eye. The images it returned, however, were nothing short of mind-boggling: the "void" was clogged with ten thousand orbs of light, each orb a galaxy like our own Milky Way, home to hundreds of billions of spiralling stars. Expanding that slice of night sky, scientists have estimated that there are at least one hundred billion galaxies in our universe, filled with a sextillion stars.[33] Think of how incredible that is: we are surrounded by 1,000,000,000,000,000,000,000 stellar giants, but they are too faint for our eyes to see.[34]

While we've been staring at the stars for millennia, it is only recently we've come to know that these twinkling pinpricks of light are actually massive nuclear reactors, balls of hot luminous gas that are blast furnaces of atomic fusion—that brought up close, even Alcor, barely visible as "the forgotten one," would dwarf our own sun and, at thirteen times its brightness, burn up our entire sky. It is almost cosmic trickery, then, that from our

---

33 Using new techniques, researchers have re-imaged the data and calculated that some galaxies may be nearly twice as big as previously thought.

34 "The faintest galaxies are one ten-billionth the brightness of what the human eye can see."

vantage point the most massive things in the universe appear to us as if the sky were a petri dish and the stars mere specks.

SIZE IS PHYSICAL, but it's also a mental construct that we grapple with as well. The problem is that our brains are not very good at processing how immensely big or small things can get once they are beyond our perceptual limits. As the English writer Helen Macdonald has observed, "We are very bad at scale. The things that live in the soil are too small to care about; climate change is too large to imagine." Instead, at imposing scales, things, objects, numbers tend to blur away into what researchers call "scale blindness." The vastness of the universe and the infinitesimal quantum world may be fundamental to our existence, but for the most part we spend our days unaware of the larger and smaller scales we inhabit.

To illustrate what I mean, take a moment and picture in your mind a single cupcake. It should be easy. Now picture ten. Keep expanding that number up to see if you can see fifty cupcakes in your mind, or a hundred. The resolution of the cupcakes will fade, but the bulk of cupcakes should still be visible. But now, widen the scope: try to picture a thousand cupcakes, or a hundred thousand. As the number gets larger, especially to a million or a billion, our ability to envision the scale, let alone the individual cupcakes, completely breaks down. This might seem like a small issue, and it is when the topic is trivial, like cupcakes, but there are bigger implications when the topic is serious.

We may live in a world of big data, but we are numb to big numbers. And the figures that are fed to us on the news each day are mostly incomprehensible. Whether it's the forty-six million acres of trees that are deforested each year,[35] the $20-trillion US

35 The number in football fields is just as mind-numbing and hard to fathom: 60,720,000 football fields.

national debt, the $1,676 billion spent annually on weapons and arms, or the twenty million people on the brink of famine and starvation, when it comes to big numbers the result is the same: our eyes glaze over and we find ourselves lost in the enormity. As Joseph Stalin is reputed to have said, "One death is a tragedy; one million is a statistic."

As a consequence, scale blindness can be monstrous, because we can't *feel* once we lose our sense of scale, and once we can't feel, we lose the ability to react appropriately. A team of US researchers examining this sense of scale wanted to look at the effects of putting a market price on the scale of damage to life. Specifically, they wanted to know what the perceived "cost" would be of rehabilitating thousands of seabirds after an oil spill.

To see how much people were willing to pay to fix the problem, the magnitude of the hypothetical disaster was increased by a factor of ten each time. The team found that whether the number of oil-slicked birds was two thousand, twenty thousand, or two hundred thousand, the financial offer of help was about the same. Meaning scale simply did not register. On average, the subjects showed a willingness to pay about $80 to help two thousand seabirds, but when the number of seabirds rose to twenty thousand, they offered to donate $78, or two dollars less, and when the number increased by a hundred times, to two hundred thousand birds, the price valuation rose to only $88; that's 198,000 more birds but only an $8 difference.

If we can be so easily confused by a scale-shift of a factor of 10, imagine the blur that occurs when we re-scale by a factor of a million. Today, our microscopes are so powerful that we can magnify objects over a hundred million times, allowing us to see and move the very building blocks of the universe: atoms.[36] Physicists know, however, that even this horizon keeps

---

36 The smallest unit we can see with an electron microscope an angstrom, which is 1×10^{-10}

shifting and that far more exists beyond the limits of what even our most advanced technologies can see. At present, what's believed to be the farthest end of the subatomic universe—at less than 0.000000009 yoctometres—is what's known as the Planck length: a space that is 10^{-35} smaller, or thirty-five orders of magnitude smaller, than our present scale, or what we consider our daily "reality." To put the scale of this tiny chasm in another way: a single hydrogen atom is ten trillion trillion Planck lengths across. Compared to the measure of a single Planck length, an atom is gargantuan.

Flipping over to the opposite end of the scale, the observable universe stretches out to 10^{26} metres, or ninety-two billion light years, away. This distance is equally unfathomable to us. To offer some perspective, one light year is just under ten trillion kilometres away. And just counting to *one* billion, let alone ninety-two, would take you over thirty years. "Common sense works fine for the universe we're used to," Carl Sagan has said, "for time scales of decades, for a space between a tenth of a millimetre and a few thousand kilometres, and for speeds much less than the speed of light. Once we leave those domains of human experience, there's no reason to expect the laws of nature to continue to obey our expectations, since our expectations are dependent on a limited set of experiences."

Our experiences tell us that reality is human-sized, but our technology tell us it's not. On the true scale of things, we are microscopic giants—massive and puny at the same time. And yet, even within this unimaginably boundless realm, we have a surprising "place." Positioned between these micro and macro realities, you are closer in scale to the farthest reaches of the known universe than you are to the Planck length.

THE NEXT TIME YOU ARE ALONE in a room, take a moment to consider that everything around you, every surface, every breath of air, every inch of your bare skin, is alive and humming with invisible life. And then remember that from up in the sky, let's say for a plane passing by overhead, you too are just an invisible speck.

Galileo's genius was in realizing that what humans could perceive was only a mere slice of reality. And while he was the first to see beyond the old world view, he was troubled by how others refused to open their eyes.[37] In 1632, in *Dialogue Concerning the Two Chief World Systems*, he wrote,

> In the long run my observations have convinced me that some men, reasoning preposterously, first establish some conclusion in their minds which, either because of its being their own or because of their having received it from some person who has their entire confidence, impresses them so deeply that one finds it impossible ever to get it out of their heads. Such arguments in support of their fixed idea as they hit upon themselves or hear set forth by others, no matter how simple and stupid these may be, gain their instant acceptance and applause. On the other hand whatever is brought forward against it, however ingenious and conclusive, they receive with disdain or with hot rage—if indeed it does not make them ill. Beside themselves with passion, some of them would not be backward even about scheming to suppress and silence their adversaries.

Galileo poked at the reality bubble, and he was punished for doing so. After the first injunction against him in 1616, he could no longer hold, defend, or teach Copernican astronomy. In 1633,

---

37 Galileo was so controversial it took 350 years for the Catholic Church to officially concede, in 1992, that he and Copernicus were right.

he was tried again, this time by the Roman Inquisition, and was found guilty. Because of his fame and old age, however, the great astronomer was spared the heretic's punishment of torture and death. Instead, Galileo was fated to spend the rest of his life under house arrest.

It is quite something when you consider that two of science's greatest men—Van Leeuwenhoek, the father of microscopy, and Galileo, the father of modern astronomy—both spent years being mocked for seeing the true nature of reality. In the end, both had the final word: Van Leeuwenhoek was recognized by the Royal Society and became eminent among his peers; Galileo is now considered one of the greatest thinkers who ever lived. But Galileo, also known as the "father of modern science," left behind more than just a scientific legacy.

On March 12, 1737, ninety-five years after his death, Galileo's grave was robbed. The thief was Anton Francesco Gori, a professor, who cut off three of Galileo's fingers when his body was being moved from a temporary tomb to the Basilica di Santa Croce in Florence. The practice of removing dead saints' fingers and other body parts was common, as the relics were believed to hold sacred powers. Gori was honouring Galileo as a martyr, a secular saint of science, who blasted apart old beliefs and liberated us with his thinking.

It was almost two centuries later, in 1927, when the first of Galileo's missing fingers was found. Today, it is on display at Florence's Galileo Museum. The symbolism, for those who know the story, is hard to miss. There, sealed in a glass jar, defiantly pointing to the heavens, is Galileo's middle finger.

2

MIND BOMB

It has never occurred to me before, but this is
truly how it is: all of us on earth walk constantly over a seething,
scarlet sea of flame, hidden below, in the belly of the earth.
We never think of it. But what if the thin crust under our
feet should turn into glass and we should suddenly
see. . . I became glass. I saw—within myself.

—YEVGENY ZAMYATIN

THE DETECTIVES AT THE SCENE didn't have much to go on. The scene was grisly. Lying on the floor of the apartment were two elderly women: one was found with her head beneath a chair, the other wrapped in a carpet. They took fingernail clippings as well as hair and skin samples from the deceased. Both women were mummified.

The Viennese police discovered the sisters' bodies in 1992. They had been dead for years. Neighbours hadn't missed them, as the women were recluses. Instead, many had assumed that they had simply packed up and moved away. The city bankers, however, were more curious than the neighbours, especially as the sisters were wealthy. So, when their accounts fell dormant, questions began to arise, and eventually the police were sent to investigate.

News of the women's demise soon spread to the insurance companies. And while foul play was not suspected, the insurers still needed to know which sister died first, as the company holding the policy for the inheritor had a large amount of money to gain. For the forensics team, this was a mystery, and they turned to physicists at the University of Vienna for help. To crack the case, the scientists would develop a new tool with the ability to "see" the body in a brand-new way: a kind of clock, one with the power to pinpoint their time of death.

To explain how the clock works, we must first go in time back to 1763. If you'd been walking along the cobblestone streets of Harwich, you might have heard a couple of brilliant minds deep in conversation one day. Samuel Johnson and James Boswell were debating the ideas of a famed Irish philosopher named George Berkeley. Berkeley's argument was quite radical for its time. He believed that we cannot know things as they truly are; instead, what we know about the world is based on our sensory impressions of things. Put another way, for Berkeley, our perceptions make things appear "real" to us; that is, a table or a chair exists only insofar as it can be perceived, meaning that these objects exist only in our minds. What troubled people back then, and continues to trouble us now, was that since we can know only the world that exists in our minds, how do we know for certain that the material world is even really there?

It seems an impossible riddle, but Johnson was of the belief that Berkeley's argument could be proven wrong, in a rather simple manner. As the story goes, he confronted the philosophical problem with a decidedly non-philosophical solution. How would he refute Berkeley's assertion? Boswell looked on as Johnson kicked a large stone, yelling, "I refute it THUS."

It was in this way that Johnson gave to the history of philosophy a new fallacy: the *argumentum ad lapidem*, or "appeal to the stone." Because, of course, he hadn't refuted Berkeley at all. The pain in Johnson's toe was exactly what Berkeley would have predicted: a pain that was only real because it had been created by his mind.

Today, this same question of external reality is being tackled not only by philosophers but also by scientists. Is there a world "out there," or does it require a consciousness for it to be perceived? What we see is certainly not what objectively exists. Instead, what we see is based on our particular human sensory machinery. As Robert Lanza and Bob Berman write in *Biocentrism*, a yellow, flickering candle flame, for instance, cannot be perceived without us:

> The flame is . . . merely a hot gas. Like any source of light, it emits photons or tiny packets of waves of electromagnetic energy . . . these invisible electromagnetic waves strike a human retina, and if (and only if) the waves each happen to measure between 400 and 700 nanometres in length from crest to crest, then their energy is just right to deliver a stimulus to the 8 million cone-shaped cells in the retina. Each in turn sends an electrical pulse to a neighbour neuron, and on up the line this goes, at 250 mph, until it reaches the warm, wet occipital lobe of the brain, in the back of the head. There a cascading complex of neurons fire from the incoming stimuli, and we subjectively perceive this experience as a yellow brightness occurring in a place we have been conditioned to call "the external world."

The same thing holds true for hard physical objects, like rocks. There is nothing solid about a rock. It is made up of a fizz of atoms and flickering subatomic particles, the bulk of which is empty space. What Johnson perceived when he kicked the rock that day was a sensation of pressure as the negatively charged electrons in the outer shell of the rock repelled the negatively charged electrons that made up the outer shell of his shoe. There was no solid contact; it was just pressure translated by his brain as sensation. From the nerves in his toe, all the way up his spinal cord and into his brain, this was how Johnson came to perceive that he kicked a stone. To date, no one has effectively refuted George Berkeley. Even Albert Einstein could not definitively prove that reality exists. In a 1955 letter he wrote, "It is basic for physics that one assumes a real world existing independently from any act of perception. But this we do not *know* [emphasis in original]."

While we can't say for certain whether reality exists independently of an observer, what we do know is that the physical world is far stranger than what our eyes perceive. For one thing, we commonly think of our bodies as separate and distinct from the external world, but modern science tells us that there is no "out there"; indeed, there is no place where your body ends and the world begins.

IF YOU KNOW HOW TO SEE, you can make an entire mountain disappear. That is exactly what a team of scientists did in the Gifu prefecture of Japan. Here, Mount Ikeno rises over the landscape. With snow-capped peaks and a river winding past its base, it looks the backdrop of a postcard.

But hidden within, a mile beneath the summit, is a high-tech lab that would rival a villain's lair in a James Bond movie. Inside this former zinc mine, technicians in hooded white coveralls monitor a thirteen-storey-tall steel tank filled with fifty thousand metric tons of ultra-pure water. This is the Super-Kamiokande, or Super-K, neutrino observatory, a massive underground facility built to detect some of the smallest known subatomic particles in the universe. To "see" these invisible particles, the observatory's ceiling and walls are lined with eleven thousand shiny hand-blown glass bulbs, called photomultiplier tubes, built to pick up tiny flashes of neutrino light.

It was here, buried deep in the heart of the mountain, that this $100-million "camera" took the most extraordinary picture of the sun. The image is pixelated but instantly familiar: a white-hot core, ringed by bright yellow, and flaring out into orange and waves of red. But what's puzzling is this: how could an image of the sun be taken from deep inside a mountain with no windows to let the light in? The answer lies in what the Super-K is looking for. While a regular camera captures photons, or particles of light, the Super-K captures and images a different kind of particle. It is looking for neutrinos: particles so tiny and fast moving they can zip right through even the densest of matter.

Imagine a block of lead stretched out one light year in length, or nine and a half trillion kilometres long.[1] Now imagine shooting a stream of neutrinos through it at one end. Incredibly, half of the particles would sail through effortlessly, making a clear exit on the other side. At nearly zero mass, with no electric charge, neutrinos are so named because they are neutral and are not drawn to other subatomic particles. They are also unimaginably puny. For a neutrino, the space between lead atoms is like an immense chasm, as a result, zipping right through lead is a breeze.

In everyday life, our eyes and stubbed toes deceive us into

---

1 A hydrogen atom has one proton, no neutrons and one electron. A lead atom is much "busier." It has eighty-two protons, eighty-two neutrons, and eighty-two electrons, which is why lead is a much denser element.

believing that matter is solid, but that is an illusion. As early as grade school, we learn that atoms are made up primarily of open space. For example, if you were to blow up a hydrogen atom's nucleus to the size of a golf ball in your hand, the rest of the electron shell around you would be a kilometre away. But to understand how small a neutrino is, you would need to make the atom enormous—you'd have to inflate it to the size of the solar system. At that scale, the golf ball in your hand would be about the size of a neutrino; that's how minuscule a neutrino is compared to the size of an atom.

But these simple illustrations don't give us the full picture. Because at the subatomic level, neutrinos don't even have a "size" per se. According to physicists, neutrinos are "point-like particles with uncertain positions"; they are less than a millionth the mass of an electron, which is why they can move freely and relatively unencumbered even through the densest of space.

At the subatomic scale, our bodies, like mountains, are also basically empty space. To neutrinos, we're like ghosts: a hundred trillion neutrinos pass unnoticed through our bodies every second as if we're not even here. But because we are bombarded by so many neutrinos—they are constantly being generated by the sun's nuclear fusion and by the destruction of supernovas—every once in a while, given their immense volume, a neutrino will hit another subatomic particle, which allows a neutrino observatory to detect it.[2] At the Super-K facility, these rare collisions are what the scientists are waiting for. When a neutrino hits an electron in the ultra-pure water, it produces a tiny blue spark of neutrino light similar to an optical sonic boom. This results in a characteristic blue glow, called Cherenkov radiation.

---

2 Sixty-five billion solar neutrinos pass through one square centimetre perpendicular to the sun per second.

After 503 days of exposure, capturing about fifteen neutrino "pixels" a day, the Super-K had collected enough of these brief neutrino flashes, which had passed not only down through the mountain but also up through the earth, that it formed the image of the sun beaming brightly outside.

The image created by the neutrinos is called a neutrinograph. And what makes the Super-K neutrinograph of the sun so remarkable is that it proves that what looks like the solid world around us, from stones to whole mountains, is indeed empty and porous. But as we'll see next, there is another scientific imaging technique that can be used to look inside ourselves, and it reveals the same thing.

Lord Londesborough, At Home,
144 Piccadilly.
A Mummy From Thebes to be unrolled at half-past Two

—formal invitation, 1850

IN THE NINETEENTH CENTURY, British high society was gripped by Egyptomania. It was a legacy of Napoleon's Egyptian campaign, which began archaeological excavations half a century earlier. Archaeologists and explorers were revealing the awe-inspiring power of an ancient empire, once as great as their own, that now lay in ruins. The desert expeditions pillaged Egypt's ancient temples and tombs, while back in Europe private collectors snapped up the ancient artifacts which had rested undisturbed for thousands of years.

As part of this fascination, wealthy collectors began hosting mummy "unrolling" parties. Guests gathered around the desiccated bodies, marvelling at the gems and amulets that were revealed as the linen bandages were theatrically unwound from the dead. The bodies were sought-after souvenirs. As French aristocrat Abbot Ferdinand de Géramb wrote in 1833, "It would be hardly respectable, on one's return from Egypt, to present oneself without a mummy in one hand and a crocodile in the other." And so the mummies were destroyed in afternoon spectacles so that the rich could entertain their friends.

Towards the end of the century, however, a new craze emerged when in 1895 a German physics professor named Wilhelm Röntgen turned the public's attention to a mind-blowing new discovery. He called them X-rays.[3] The mysterious rays could penetrate solid matter, allowing people to see through human skin right down to the bone.

It was extraordinary. The images baffled Victorian society. People were so excited by the technology that, as one author has noted, X-rays became like the iPhones of the 1890s. Soon this new form of "super-vision" was everywhere. And for the mummies at least, the X-ray images, or radiographs, offered some reprieve. It was only months after the X-ray's debut that physicist Walter König scanned the mummy of an Egyptian child, pioneering a non-invasive way to examine human remains, preserving them for posterity.

X-rays weren't only used on the dead, of course. Physicians in particular were quick to pick up on their benefits. If in the past determining the exact location of broken bones relied solely on a doctor's best guess, now, with X-ray vision, problem areas could be identified before an operation. This ability to see through flesh became especially useful at the battlefront, as medics could spot the exact location of bullets and shrapnel lodged in the bodies of injured soldiers.

---

3 Wilhelm Röntgen refused to take out any patents on X-ray technology. He believed that people should be free to benefit from his work.

The power of X-ray vision was also not limited to scientific or medical domains. It became hugely popular with the general public as well. At fun fairs and carnivals, "bone portraits" became a new attraction, and people lined up to see the shocking sight of their skeletons for the first time. X-rays also revealed the skeletal deformities prevalent in the Victorian era. For women dressed in the fashions of the day, X-rays revealed that a lifetime spent bound in tight hourglass corsets had bent their ribs and crushed their organs.

But the new trend did not only reveal the side effects of the beauty industry, it caused them too. The British entrepreneur Max Kaiser developed what he called the Tricho system for hair removal. By 1925, he had expanded the business and set up shop in over seventy-five studios across the United States. Women coming in for hair removal treatment on their upper lips were subjected to up to twenty doses of radiation.

As with any boom, the X-ray business was for a while a free-for-all, and anyone, whether they were a builder, pharmacist, or wine dealer, could open up their own lab and be considered competent enough to read a radiograph. The technology became so prevalent that until the 1950s shoe-fitting "fluoroscopes" were available at most department stores, casually offered up to shoppers who wanted to see their feet inside a pair of shoes to ensure they had a perfect fit. But it had not gone unnoticed that the powerful rays had side effects. The X-ray craze began to fall out of favour as more and more reports came in of *unwanted* hair loss, blisters, swelling, and burns, as well as cancer and even death. The irony being that despite being able to "see" better than ever before, we could not see the damage until it was too late.

Radiation, scientists were discovering, is not all the same: there are different rays with different penetrating effects. Alpha radiation, for instance, is fairly weak and can be blocked by something as simple as an outstretched hand. Alpha rays cannot even penetrate the cells on the outer layer of your skin. Because of this, in cancer treatment, alpha radiation in the form of radium-223 is commonly used to destroy cancerous masses. Inserted into the tumour, the alpha particles kill the cancerous cells, but because they can't penetrate very far, the healthy surrounding cells are left untouched.

Beta radiation, on the other hand, goes a little farther. Emitting particles of a smaller mass, this form of radiation can penetrate a few centimetres into the human body but can be stopped by a relatively "solid" sheet of plastic or aluminum. Radioactive carbon-14 in the atmosphere is a form of beta radiation that barely penetrates the most outer layer of dead skin on our bodies. But as we'll see shortly, this form of beta radiation has other clever ways to make its way in.

As for gamma rays and X-rays, these two types of radiation have the highest penetration; they are able to travel right through the body as if it isn't even there. But they can't filter through materials the way neutrinos can. Remember, a neutrino can travel unimpeded through nine and a half trillion kilometres of lead, while an X-ray will be stopped by a few centimetres. Still, the thickness and density of the calcium in our bones are enough to block X-rays, which is what creates the images of our skeletons. Soft tissues, like our fat, muscles, and skin, are more permeable, whereas materials made up of higher-atomic-numbered elements, like calcium, or bullets made of lead, block most of the X-ray beam, creating that now familiar white silhouette.

As for our cells, for the most part X-rays shoot through them without damage, but as ionized radiation with enough energy to knock out an atom's electrons, every once in a while they can rip a cell's molecular structure, causing a mutation in the DNA.[4] That's why large or frequent X-ray doses are dangerous, because they literally shower the cells with radiation. Like playing Russian

4 DNA itself was first imaged using X-rays.

roulette, with each shot you are increasing the odds of a damaging effect.

You may have noticed that to protect radiologists and medical staff, the doors and walls of modern hospital X-ray rooms are lined with lead, and that during exams, patients must wear a lead vest to cover the parts of their body not being X-rayed. The expectation is not that *none* of the photons will make it through—some will—but with a high atomic number, the lead shield will block the vast majority of them.

Similarly, the metal luggage scanning boxes at airport security are lined with lead. Because X-rays light up high-density objects, this alerts screeners to the potential presence of weapons or bombs. If you've ever wondered why passengers have to remove their laptops and cameras from their carry-on luggage at security, it's because the X-rays can't see through these impenetrable materials, making it difficult to detect objects that might be hidden behind them.

It's been more than a century since the discovery of X-rays, and today most of us take it for granted that X-ray machines allow us to see what we were once blind to. But in a way, each new form of sight reveals a different kind of blindness. Just as the X-ray machine can spot contraband but is blind to the baggage, the Super-K can see the sun but is blind to the mountain. Sometimes, to see one previously hidden thing, we lose sight of something else.

At the heart of Johnson and Boswell's debate back in 1763 was whether there is a real distinction between what is inside our minds and what is out there in the physical world. At the human scale, we tend to see "solid" and "real" as the same thing, but at the subatomic scale the world around us is in a constant dance of interchanging particles. What modern scientific tools have revealed to us is that not only is there no strong distinction between us and the stuff around us, our bodies are made *of* the stuff around us. That stone and Johnson's toe, as we will soon discover, both have their origins in the same thing.

IN 1957, A SCIENTIFIC PAPER now known as the B2FH paper forever changed how we see ourselves on Earth. "B2FH" stood for the last names of its authors: astronomers Geoffrey and Margaret Burbidge, William Fowler, and Fred Hoyle. In it, they outlined the "stardust" theory of the origins of the living universe. And today, most of us have solid proof that all life, and all matter that makes up the stuff of our material reality is derived from elements created by the stars.

In technical terms, it's called stellar nucleosynthesis. What it means is that all of us are the physical resurrection of dead stars. That's because every life on Earth, every *body*, is born from a galaxy of explosions. According to NASA astronomer Michelle Thaller, the iron that makes our blood red was made in the final moments before a star died. For all of us, then, our very lifeblood began with a spectacular death in a solar system.

Stars themselves are born in molecular nurseries. As gas clouds made up mostly of hydrogen spiral inward together under the force of gravity, the hydrogen atoms begin to fuse together in a blinding hot core. The fusion of four hydrogen nuclei creates a new element, helium, and it is this blasting outflow of energy generated from the massive nuclear reaction that supports the star, preventing it from collapsing inward under the pressure of its own weight.[5] A star remains stable as long as these two opposing forces—exploding outward and crushing in—balance each other out.

Eventually, however, the hydrogen fuel runs out and the star begins using the only other fuel it has available: the shell of helium it produced from nuclear fusion. Fusing three helium atoms together, it begins to form the next element: carbon. The carbon then forms oxygen, and the oxygen turns into silicon and sulphur. This process of lighter elements fusing to form

5 Our own sun fuses approximately 620 million metric tons of hydrogen into helium every second.

heavier elements is the chain reaction of "stellar nucelo-synthesis,"[6] which continues all the way up the periodic table until the star reaches iron, whereupon the star becomes so heavy that the energy no longer burns outward and instead all of that power is absorbed.[7] The result is an explosion so massive and spectacularly violent that the dying star burns brighter than all the other stars in its home galaxy combined. This is the legendary supernova. It is from this stellar explosion that the primary elements in the periodic table are made: the carbon in our bodies, the silicon in our cell phones, the uranium we use to make bombs and power cities. Almost all of the matter that surrounds us came from the death of a star.[8]

Supernovas are so powerful, they can even eject atoms into other galaxies. The process is known as "intergalactic transfer," and astrophysicists at Northwestern University have calculated that approximately half of the matter that makes up our bodies is not even from the Milky Way. Atomically, Earthlings are extragalactic beings, as half of the particles in our bodies were born in far-flung galaxies. As astrobiologist Caleb Scharf writes in *The Zoomable Universe*, "In simple terms, we all condensed. The fundamental physical properties of the universe conspired to pull together a set of atoms and molecules that previously had been occupying a volume a billion trillion times larger. . . . Five billion years ago, your atoms were about ten

---

6 The astute reader will have noticed that we skipped from helium to carbon—and that the lighter elements of lithium, beryllium, and boron should be in between. These elements are cosmically created in a different fashion, when a heavier element is hit by cosmic rays.

7 Iron cannot release energy by fusion because it requires a larger input of energy than it releases.

8 Some elements, like gold, are made from the explosive collision of neutron stars.

million times more widely spread across the cosmos than they are now."

And some of these atoms are as old as the Big Bang itself. In fact, 98 percent of the hydrogen atoms in your body date back to the universe's beginnings.

The molecules that surround us are ancient too. We like to think the water we drink is fresh, but scientists believe that water is older than the sun. When you next take a sip, take a moment to consider that the water you are drinking has been a cloud, an iceberg, and a wave, that it has drifted and meandered through canyons at the bottom of the sea. Before entering your body, it has spent, on average, three thousand years in the ocean and just over a week in the sky before falling as rain. Locked up in glaciers, it rests for longer, from thousands to hundreds of thousands of years. Then one day it finally melts, spending half a month in streams and rivers before draining back out to the sea. And this cycle has repeated many, many times in the four and half billion years that Earth has been orbiting our modest sun.

It's not just water that gets recycled. The majority of the carbon that makes up our bodies, approximately two-thirds, comes from the plants that we eat and from the carbon dioxide that they exhale, but the remaining one-third comes from carbon that was trapped in buried oil and gas deposits for hundreds of millions of years. As we burn up these fossil fuels, they release into the atmosphere the carbon atoms that made up the bodies of the first aquatic animals that existed 500 to 600 million years ago; the first land plants of 475 million years ago; the earliest reptiles, insects, and amphibians of 350 to 400 million years ago; and the dinosaurs that roamed as giants from 230 to 65 million years ago. So in some small sense, you are the atomic resurrection of a dinosaur.

What this means is that while your body is constantly renewing itself, creating millions of new cells every second, the atomic materials from which those cells are made are as old as time. Like microscopic Lego, the atoms that have been used to build your body have been used billions of times before, and the atoms that are in your body right now will be used billions of times again.

At an intuitive level, we all know that life is a cycle, that "from ashes to ashes, dust to dust," the nursery for new life is a literal deathbed of rot and decay. But scientists can now see this resurrection as it unfolds. At Sheffield Hallam University, Malcolm Clench, a professor of mass spectrometry, became the first to track atoms as they moved from an organism after its death to become visibly incorporated in the body of a new life.

Producers at the BBC contacted Clench, as they were working on a documentary about the science of decay and wanted to find a visual way to show viewers the death-to-life process as it was taking place. So, Clench created an "After Life" garden, growing hydroponic plants which he fed with a special nutrient system containing chemically labelled nitrogen-15. Nitrogen is essential to life because it's a fundamental building block of our DNA. And while nitrogen-14 is everywhere in the air and very common, nitrogen-15 is exceptionally rare and only 0.3 percent naturally abundant; that is, you're unlikely to encounter it by accident.

Clench's donor plants were grown to be sacrificed. After mulching them, he turned the dead matter into a liquid compost. This "death soup" was fed to new seedlings which until that time had been grown with the abundant isotope of nitrogen-14. Then, using a mass spectrometer,[9] which sorts and isolates atoms and other

9 MALDI-MS is a mass spectrometry system that uses a laser to sort atoms and other compounds, allowing scientists to see what an object is made of by looking at their mass and charge. Modern instruments charge an atom or molecule and the laser serves as a firing gun, starting a literal atomic race. The lightest ions are the fastest, and the heaviest the slowest. And so, based on their speed and atomic weight, you can develop a picture of the compounds present in a sample.

compounds by mass, Clench was able to generate a photo of the nitrogen's exact location, showing where it had been taken up in the young plants' leaves. It was possible to see with special imaging tools that the nitrogen-15 in the leaves lit up and glowed a bright white. The atomic "death marks" of the rare isotopes could only have come from one place: it was the atomic resurrection of the dead donor plant.

Life and death are a cycle. That's just the way nature works. In Alaska's Tongass rainforest, a similar process is visible. But here, what scientists are looking for are salmon in the trees. Typically, we picture animals eating plants, but in this instance it's the trees that are feeding on animal remains.

Each year, when hundreds of millions of salmon return to the rivers and streams to spawn, they die and decay there, becoming chemical nutrients for the forest. As biologist Anne Post notes, a spawning chum salmon contains an average of 130 grams of nitrogen, 20 grams of phosphorus, and more than 20,000 kilojoules of energy in the form of protein and fat. That means that in just one month, a 250-metre-long stream where salmon come to spawn and die receives more than 80 kilograms of nitrogen and 11 kilograms of phosphorus.

Because of this, the Tongass is known as a "salmon forest." Scientists tracking stream-side vegetation have found that anywhere from a quarter to three-quarters of the trees' nitrogen comes from returning salmon. This can make a huge difference to their growth. Sitka spruce growing by these riverbanks take around eighty years to reach fifty centimetres in trunk diameter; their salmon-less counterparts in the interior require much longer, on average, three centuries to reach that size.

The sitkas' tree rings also show a record of the salmon's return. In years where the salmon runs have been large, the marine-derived nitrogen-15 in the trees' sap is highly correlated. As you may remember, nitrogen-15 is very rare in terrestrial environments, but it is common in the marine food web. The nitrogen-15 in the trees could only have come from one place: the returning fish. Meaning that the salmon's spawning history is literally being written in the library of the forest.

Humans are not exempt. We too are subject to the same death-to-life process. Though we may squirm to think of it, naturally buried human bodies also enrich the soil, and just like the salmon we leave our chemical signatures behind. After death, for every kilogram of dry body mass, the average human body releases thirty-two grams of nitrogen, ten grams of phosphorus, four grams of potassium, and one gram of magnesium into the soil of a gravesite. And while a burial will initially kill off some of the nearby vegetation, eventually a balance returns and our decomposing bodies begin to nourish the ecosystem.[10] Just as dying stars gave rise to life on Earth, our own scattered atomic remains re-form in new bodies. They become the ingredients of life again.

Since the Big Bang, there has been no new matter in the universe, but over the past hundred years or so, scientists have discovered how to transform atoms in ways that are highly unlikely to occur in nature—some of them intentional, some less so. Of course, even something as familiar as burning a piece of toast, or baking bread in the first place, means changing molecular structure. (Which is pretty impressive, from an evolutionary perspective: one way of thinking of about human ingenuity is that we are an animal capable of modifying molecular structure.) But that's not nearly the same thing as making new elements, the way a star does. That requires unimaginable amounts of energy, something far beyond what our ancestors could have dreamed of.

---

10 Not with embalming fluids or cremation though. Those are bad for both the soil and plants.

Today though, we have that power. Of the 118 elements on the periodic table, 26 are synthetic, or human made. We make new elements by smashing atomic nuclei together in a process called fusion. The particles are made to collide at high speeds in a particle accelerator, fusing them into a heavier element.

We have also developed the opposite superpower: the ability not only to fuse atoms but also to make them split. And we demonstrated this power at 5:29 A.M. on July 16, 1945. A photograph captured by the United States Department of Defense shows a bomb codenamed Trinity, a half-second after detonation: in it a dome three hundred metres wide, rises up over New Mexico's Jornada del Muerto desert like a giant blister. Inside, a fireball ten thousand times hotter than the sun is set to explode into a deadly mushroom cloud.

Trinity was the world's first nuclear weapon. The power of the bomb unleashed the energy equivalent of 20,000 tons of TNT. As it did, smoke and debris erupted 11,600 metres into the sky, causing a downpour of radioactive confetti. At the surface, the shockwave blasted a crater into the earth, the heat liquefied the sand, and even 16 kilometres away, observers felt like they were "standing directly in front of a roaring fireplace." For the first time, humanity held in its hands a power as awesome as the sun. And by daybreak, within a 1.5-kilometre radius of the test site, nothing remained alive.

Once the United States unveiled this weapon of mass destruction, it was only a matter of time before everyone else wanted it too. Over the next two decades, the world's most advanced militaries raced to build their own bombs, pockmarking the planet with over five hundred white-hot explosions and sending tons of radioactive debris up into the atmosphere until the Limited Test Ban Treaty was signed in 1963. Then things quieted down. Nuclear tests were banned in outer space, underwater, and in the atmosphere, but no one was aware that after detonation the remnants from nuclear bombs don't just vaporize and disappear. Each blast injected radioactive particles into the atmosphere, rushing the molecules towards a new destiny. Just like exploding stars, the bomb blasts would become new life.

For life to exist, however, the element of carbon is vital. All life on Earth is made of it. The same stuff that makes us up is also found in lumps of coal, pencil lead, and diamonds. In living things, it's a primary element found in proteins, sugars, fats, muscle tissue, and DNA.[11] Plants inhale it directly from the atmosphere, and animals in turn absorb it from the plants they eat. For humans, like all plants and animals, the carbon we take in is used to build our bodies, which brings us back to the mystery of the two mummified sisters in Austria.

THE FORENSICS TEAM INVESTIGATING the sisters' death approached nuclear physicists for help because they knew that radiocarbon dating was used to establish the age of Egyptian mummies. But natural atmospheric carbon-14, which is typically used in carbon dating, was useless here[12]; with a half-life of 5,730 years, carbon-14 could certainly be used to date organic tissue, but the accuracy range was in the order of several hundred years. To find out which of the sisters died first, the physicists needed a measure on a human time scale.

A light bulb went off when they realized there might be a different way to date the bodies. They could look for the artificial spike of carbon-14 that resulted from the fallout of Cold War nuclear testing. Combined with oxygen, carbon-14 becomes

---

11 When DNA replicates, 30 percent is carbon.

12 The primary source of C^{14} is cosmic ray collisions.

carbon dioxide, which is inhaled by plants. When animals eat plants, or eat other animals that feed on plants, they take in that carbon. And because cells don't discriminate, these carbon isotopes work their way into the food chain. That's how radiocarbon from nuclear bomb blasts became a building block of every living being.

Carbon-14 is rare in the environment. It only makes up one-trillionth of the carbon on the planet. This special spike in radiocarbon can be detected in us because the amount of carbon-14 in the atmosphere doubled during the era of above-ground tests before abruptly plummeting again after the test ban. For physicists, this bomb-pulse curve can be read like an atomic calendar, because since that time the radiocarbon has diluted at a steady rate of 1 percent a year. If the scientists could measure the amount of "artificial radiocarbon" inside a cell, like a time-stamp, they could pinpoint the date that the cell appeared.[13]

The forensics team now had a way to unravel the mystery. What they needed next was a sample of cells from the sisters' bodies that regenerated quickly, cells that were created in days or months rather than years.

You may have heard the myth that every seven years all of your body's cells have been replaced so essentially, you've become a brand new person. And while it is true that you lose, on average, about fifty billion cells a day, the cells in our bodies have vastly different lifespans and turn over at different rates. Some cells are like mayflies and die within a few days, while others are programmed to stay on with us for weeks, years, or even decades.

---

13 As the bomb-pulse radiocarbon decreases by 1 percent per year, by 2030 the bomb pulse will die out. That's because organisms born after this time will no longer have any significant spikes of the bomb pulse traces, and so their cells will not be able to be timed. That is, unless we set off more bombs.

And to fully bust the myth: there are some cells so loyal, they stay with us our entire lifetimes.

Skin cells are quick to go. Stationed on the front lines of our bodies, these cells are replaced every two to three weeks. The entire outer layer of our skin, the epidermis, is exchanged every second month or so. But not only the external parts of us are quickly renewed. Deep inside our guts, our intestinal cells known as villi have even shorter lifespans. Exposed to grinding stomach acids, they go through tremendous wear and tear, shedding and regenerating every couple of days. The speed at which cell turnover takes place also depends on the cells' vulnerability. While the surfaces of our corneas come with the added protection of our eyelids, these cells are vital for focused vision and so we have a built-in emergency response, and if any damage occurs we can replace them in as little as twenty-four hours.

Joining us for a longer ride are the cells in our bones. Our skeletons are broken down and gradually replaced every decade. Our heart cells stay with us even longer. In our twenties, we replace them at a rate of 1 percent a year, but this regeneration slows down, and we replace less than 0.5 percent of heart cells annually by the time we're seventy-five. So, if you live to the ripe old age of one hundred, you will still have about half of the original heart you were born with.

Because no new carbon is taken in after a body is dead, by examining skin and hair samples, some of the last new cells the sisters' bodies had made, the scientists determined that one sister had died a year earlier than the other, in 1988. Her cells contained more carbon-14 from the bomb pulse. The last cells were formed in the other sister's body in 1989, meaning she must have lived alongside her dead sister's decomposing body before dying herself the following year.

WHERE DO WE END and where do we begin? As children, the answer seems simple: I am "me," and everything else is separate. In fact, even infants have an intuitive understanding of physics. They understand, for instance, the notion of solids: that two solid objects cannot occupy the same space, and that most objects are persistent and have stable boundaries. This is common sense to most of us from our earliest years, but it is a natural blind spot. Given the scale we inhabit, we perceive as solid what is in fact porous, and what seems separate from our bodies is, at the atomic and subatomic levels, deeply interconnected to everything.

The mystics have long understood this. As Lanza and Berman note, "Entire religions (three of the four branches of Buddhism, Zen, and the mainstream Advaita Vedānta sect of Hinduism, for example) are dedicated to proving that a separate independent self, isolated from the vast bulk of the cosmos, is a fundamentally illusory sensation." In the practice of Zen Buddhism, the aim is to make the invisible visible. Much like science, the goal is for the Zen practitioner to realize that there is "no separation between the self and the ten thousand things." The famed Buddhist monk Thich Nhat Hanh illustrates the idea in non-scientific terms by describing a simple flower. A flower, he explains, cannot exist as an isolated thing, because it is intimately connected to everything around it:

> Looking into a flower, you can see that the flower is made of many elements that we can call non-flower elements. When you touch the flower, you touch the cloud. You cannot remove the cloud from the flower, because if you could remove the cloud from the flower, the flower would collapse right away.
>
> You don't have to be a poet in order to see a cloud floating in the flower, but you know very well that without the clouds there would be no rain and no water for the flower to grow. So cloud

> is part of flower, and if you send the element cloud back to the sky, there will be no flower. Cloud is a non-flower element. And the sunshine . . . you can touch the sunshine here. If you send back the element sunshine, the flower will vanish. And sunshine is another non-flower element.
>
> And earth, and gardener . . . if you continue, you will see a multitude of non-flower elements in the flower. In fact, a flower is made only with non-flower elements. It does not have a separate self.

All living things are like this. We are not isolated; we are networks. Life could not exist otherwise. We are made of matter, and like all matter, we are bound by the second law of thermodynamics, which states that an isolated system will always tend towards a state of chaos and disorder. As living systems, as *organized* matter, we fight this entropy through constant inflow from the outside world. And we can do this because living things are not closed systems. We require energy from the world around us to maintain our existence. In a very real way, what we call death is the moment that this exchange stops and we dissolve back into chaos. We lose our solidity and become particles again.

George Berkeley's question of whether the material world is "real" or only an impression of the mind presupposed that our minds are made from some other "stuff." Today, we know that our brains—our minds—are made of the very same primordial elements that we now observe. Our second blind spot is that we cannot see how intimately connected we are to the universe around us. That in reality, as astronomer Michelle Thaller has observed, "We are dead stars, looking back up at the sky."

3
I TO EYE

When one does not see what one does not see,
one does not even see that one is blind.

—PAUL VEYNE

FOR GÉZA TELEKI, it was a rare day off. The primatologist had set out for a scenic hike along one of the high ridges in Tanzania's Gombe National Park. Towards late afternoon, he found a perfect spot looking out over lush rolling grasslands. There, he settled under a tree to await the evening spectacle: the big African sun would soon drop over the glittering waters of Lake Tanganyika.

It was quiet above the valley forests, but as Teleki looked around, he realized that he was not alone. Climbing up from opposite directions were two adult male chimpanzees. As they reached the ridge crest, they spotted one another. They both got up on their hind legs, walked upright through the grass until they met eye to eye, and greeted each other by softly panting and clasping hands. Now just a few yards in front of Teleki, the chimpanzees sat down. All three sat together in silence. For the primatologist, it was an experience that was transformative and profound. The chimpanzees had come to the spot, just as he had, simply to sit and watch the beautiful sunset.

What are we to make of this? Given that we are 99 percent chimpanzee, that we share largely the same DNA, is it really so impossible that they could appreciate something like a sunset? Or is that anthropomorphism? Are we projecting our thoughts and ideas onto another species, seeing the chimps' behaviour through a human lens?

There are at least two ways of looking at it, and both reveal that we have a blind spot with regard to how we view other species. On the one hand, we have to concede that we are not, as we might think, the planet's only stargazers. Indeed, we are not the only problem-solvers, not the only communicators, and not the only animals capable of love or the appreciation of beauty.

But the other way of looking at the chimps' behaviour may be even more astonishing, because, though we can guess at the thoughts or emotions of our fellow primates on that hillside, the truth is their experience is completely unknowable to us. That is, even our closest evolutionary relative might see and perceive a world completely different from our own.

Most of us spend little time thinking about how other animals perceive the world. But in Italy, a by-law came into effect in the city of Monza that made it illegal to keep a pet goldfish in a bowl. The ruling came about because the fish have good vision, and so keeping them in a warped environment that forces them to live with a "distorted view of reality" is considered cruel. City councillor Monica Cirinnà, quoted in the newspaper *Il Messaggero*, said, "The civilization of a city can be measured by this"—"this" being the still rather shocking idea that we could or should have respect for an animal's perspective.

Goldfish do indeed have remarkable eyesight. They not only have red, green, and blue cones for colour vision, as we do, but they have an additional fourth receptor, for UV light, meaning an entirely other way of seeing is open to them that is closed to us. When you stop to think about it, it's perhaps not surprising that animals have good eyesight: we expect that they would in order to survive. But what is surprising is the *kind* of information that some animals are able to perceive.

Archerfish, for instance, can distinguish between individual human faces. The fish have a rather unique skill for an aquatic species: like biological water pistols, they spit out jets of water to shoot down aerial prey. The fish can target an insect above the water and strike accurately at even sixty centimetres away. This special ability gave researchers at the University of Oxford and the University of Queensland an idea: they wanted to see if the fish's accuracy and keen eyesight could be used in another way. So they gave the fish paired images of human faces along with a food reward and trained them to use their jets to strike at the image of one particular human face on a computer screen.

Given how similar human faces look—with the same basic structure of eyes, nose, and mouth—even we sometimes have difficulty discriminating within our own species. But for a fish, especially one with such a small brain, and one that did not evolve human facial recognition abilities, the results were stunning. Presented with a sequence of forty-four new faces paired each time with another face they had been trained to remember, the archerfish demonstrated excellent visual recognition, selecting the correct face during the trials with 86 percent accuracy.[1] If that doesn't seem impressive, ask yourself if you could pick out the face of one archerfish from a school of forty-four.

Common pigeons are also known to have highly sophisticated vision. They can distinguish each letter of the alphabet, recognize dozens of words, tell the difference between paintings by Monet and Picasso, and even recall up to 1,800 individual images. Researchers, aware of pigeons' excellent discriminative powers, wanted to see how they would fare in a highly complex task: discerning the difference between malignant and benign growths in breast biopsies. Malignant tumours that turn into cancer are often signalled by micro-calcifications in the breast tissue and are distributed in a particular way. For radiologists and pathologists, it can take years to acquire the skills to distinguish between malignant and benign masses. The pigeons weren't given years. They were trained for only thirty-four days using a touch screen attached to a food pellet dispenser.

During training, the birds were shown images on a screen and were rewarded when they correctly tapped either a yellow bar when the biopsy was benign or a blue bar to signal that it was malignant. The pigeons were amazingly accurate, making a correct identification, even on new images, at a rate of 85 percent. When the researchers took a "flock sourcing" approach and pooled the responses of all sixteen trained birds, the accuracy rate went up even higher. Together, the pigeons gave an accurate diagnosis 99 percent of the time.

The point is not that we should replace radiologists with pigeons, but we should at least begin to question our ideas of

---

1 The study has since been updated to include 3D renderings of faces. It was found that "the fish were able to continue to recognize that image even when the face was rotated by 30, 60, and 90 degrees, from a frontal view to a profile."

what intelligence means. We set the bar at the default position of human intelligence, but since we can't assume that pigeons are smarter than radiologists, clearly we have to reappraise what intelligence is.

One aspect of intelligence is our ability to interpret visual information, to make sense of and respond to the world before us. For humans, that includes the ability to perceive spatial information, to read words and decode maps, and to understand symbols. Sight is not a requirement for any of those skills, of course, but it is a sense we rely heavily upon and is an aid in navigating through our environment. And yet even without all that sophistication, a homing pigeon can do something most of us cannot: randomly dropped off hundreds of kilometres away from its loft, the bird will always—remarkably—find its way home.

Today, of course, we have GPS, but imagine if we had the inbuilt capabilities of a migrating bird or homing pigeon? We now know that fish, birds, turtles, mammals, insects, and even bacteria can detect magnetic fields. If we could do the same, how would it change the way we think? And would it make us more "intelligent"?

These are rhetorical questions, but they point to the fact that we see our world through the tiniest pinhole of perception. There are at least 8.7 million other animal species on Earth, each with its own way of perceiving. So let us take a look through some of these lenses and see how other species experience our world.

REALITY IS LIKE AN IMAGE composed of billions of different pixels, each with its own distinct view. As primatologist Frans de Waal has noted, "This is what makes the elephant, the bat, the dolphin, the octopus, and the star-nosed mole so intriguing. They have senses that we either don't have, or that we have in a much less developed form, making the way they relate to their environment impossible for us to fathom. They construct their own realities."

Meaning, what we know as "reality" is only a fractional view. Our eyesight, for example, is limited to a mere 0.0035 percent of the electromagnetic spectrum. What we call "visible light" are the wavelengths within the range of 380 to 700 nanometres. Light with a wavelength of around 700 nanometres is red; at 600 it's yellow-orange; at 500 it's green; at 400 it's blue-violet. The spectrum above and below that range is invisible to our eyes. But even what we do perceive as "colours" do not really exist in the outside world. They are interpreted inside our brains and are dependent on the number and type of receptor cells in our eyes that are attuned to particular wavelengths.

In a mist of sunshine and rain we see the brief biological wonder of the rainbow, the part of the arc that corresponds to the visible spectrum. On either side of it are the invisible wavelengths that we are not biologically equipped to see. As Philip Morrison writes in the foreword of *Super Vision*, "Go in one direction from the visible portion of the electromagnetic spectrum, and the last bit of violet fades out, giving way to ultraviolet colors, then to X-ray colors, then to ever more exotic invisible colors known as gamma rays. Go the other way and the last bit of red gives way to infrared colors, which we feel as heat instead of see as beyond-red colors. Continue in that direction and you get to the longer wavelengths that now fill the airwaves conveying radio and television programs, billions of cell-phone conversations . . . radar signals from air-traffic control towers and air-defence systems."

In other words, we think of X-rays as being invisible, but all that means is they are invisible to *us*. That is, invisibility describes not the X-ray but our own way of seeing or not seeing. Some animals perceive light in wider ranges of the spectrum than we can; specifically, ultraviolet and infrared. Snakes like pythons, boas, and pit vipers have a specialized "pit organ" between their eyes and nostrils allowing them to see in the 750-nanometre to 1-millimetre infrared range. Even blindfolded, the snakes can accurately strike their prey. That's because the pit organ is sensitive to radiant heat, picking up individual temperature readings which it uses to generate an image in the brain. In this way, a viper can "see" a warm-blooded mouse in the dark.

Bees also see beyond the visible spectrum. A black-eyed Susan flower, for instance, might appear to us as a bloom of yellow petals, but to a bee that can see down to the 300 nanometre range in ultraviolet,[2] it's lit up like a landing strip. Gardens are filled with these secret bull's eyes, invisible to us but lit up for the bees to find nectar. Golden eagles likewise see ultraviolet light—they use it to follow the fluorescing urine trails that lead to their prey[3]—but they also have killer visual acuity. While we gauge good eyesight as being 20/20,[4] eagles have 20/5 vision, meaning that an object visible to you at five feet away would be visible to an eagle from twenty feet. That's because the fovea in an eagle's eye—the part of the eye responsible for visual acuity—is much deeper than our own, allowing it to see close up, like a telephoto lens on a camera.

---

2 While previous studies have made this suggestion, a new study scrutinizes the finding. For now, more research is required.

3 Bees can see in the 600 to 300 nm range. How do we know what a bee can see? "We can find out whether an animal can see light of a particular wavelength by testing whether that light will travel through the lens of its eye. The lenses of healthy humans block ultraviolet light, so we cannot see it. But for other species, seeing ultraviolet can make it easier to see in dim light."

4 20/200 vision is legal blindness. A person with 20/20 vision would be able to read the big letter E on the Snellen chart from 200 feet away, whereas a person with 20/200 vision can see it at 20 feet.

An eagle's eyes are so good, it can spot a rabbit from 1.6 kilometres away. That's like you being able to see an ant from the top of a ten-storey building, or having nosebleed seats at a stadium rock concert but still being able to clearly see the performers' faces. Raptors also have exceptional colour vision. An eagle's fovea is packed with cone cells, giving it incredibly vibrant resolution. While humans have about two hundred thousand cones per millimetre at the centre of the fovea, eagles have one million. That's like seeing the world on an old TV set, in low resolution, versus seeing it in ultra-high definition.[5]

Humans are also somewhat limited by the placement of our eyes in the front of our heads, which gives us a field of view that's around 180 degrees. Eagles, whose eyes angle thirty degrees back from the midline of their faces, have a visual field that's 340 degrees. But while we hear the term "eagle eyes" quite often, in this particular domain hammerhead sharks have eagles beat. With their wide heads, the powerful predators have full 360-degree stereo vision. Not only can they see both in front of them and behind, these animals can also simultaneously see what's above them and below.

Even our planet's "lowliest" creatures have abilities we are only now starting to appreciate. The humble dung beetle makes its living by rolling fresh feces into a ball two to three times its own size. Then, by manoeuvring onto its front legs like it's doing a handstand, this hard-working insect uses its hind legs to push its prize backward away from the dung heap and the competition as quickly as it can.

But how does it know where to go? Face down, with a big

---

5 That said, comparatively speaking, on tests of human visual acuity, we can see detail very well compared to most species. Researchers who studied six hundred animal species found that human sight is about seven times sharper than a cat's, forty to sixty sharper than that of a rat or a goldfish, and hundreds of times sharper than a fly's or a mosquito's.

ball of feces blocking its view, the dung beetle still has an uncanny sense of direction. Scientists discovered that the beetles know where they are and where they are going by mapping the skies. If you watch a dung beetle, you'll notice it climb on top of the dung ball every so often and perform what appears to be a little dance. It has been known for some time that what they are actually doing is taking a mental snapshot, a 360-degree panorama of the sky. By comparing a mental image of the location of the sun or moon overhead to their internal map of the heavens, they are able to track their position and move continuously in a straight line.

But researchers were curious: What about moonless nights? How does the nocturnal dung beetle species get around without a bright marker in the sky? To find out, they took their tests indoors to a planetarium, where they had full control over the celestial environment. Surprisingly, when they dimmed the moon, the beetles still kept right on track. Only one other source of light remained to serve as their guide: it appeared the beetles were navigating by looking up at the Milky Way.

To be sure that this was really what was happening, the scientists needed to test the insects and limit their conditions. So they made the beetles wear little cardboard hats. This way, the researchers could see if it was in fact starlight and not some other sense that was guiding the beetles. Beetles in a control group were given clear plastic visors and could still see above them. The results were conclusive: the dung beetles wearing hats became disoriented and were unable to track their whereabouts; they rolled their dung balls around aimlessly. The control group rolled their balls almost perfectly straight ahead. These tiny Earthlings were using a distant galaxy as a compass.[6]

---

6 This intelligence, scientists have proposed, could be adopted in creating algorithms for robots or autonomous cars, a way for machines to keep track of their whereabouts, without human input or interference.

The animal kingdom is full of marvels, but when it comes to the eyesight champion? A top contender has to be the dragonfly. The speed demons have twenty-eight thousand lenses per compound eye, which together make up the bulk of their heads. They also have unparalleled colour vision. While humans are trichromatic—we have three light-sensitive proteins, called "opsins," that absorb red, blue, and green wavelengths, giving us the ability to mix as many as one million colours—some dragonfly species have up to thirty pigment opsins, allowing them to create a vast palette of literally unimaginable colours.[7] The insects can also see in ultraviolet and detect polarized light. In addition to all of that, they have another spectacular skill: they can see in slow motion.

To a dragonfly, like Neo in the film *The Matrix*, fast-moving bullets would appear slowed down, and what would appear as a speedy blur to us would be a crisp image. That's because we see at about fifty frames per second while dragonflies see at three hundred frames per second. What looks like a movie to us would appear as a slide show to a dragonfly. Which is why it should come as no surprise that the insects are such formidable hunters; with their super-vision, they are able to catch 95 percent of their prey.

We will never completely know the world of wonder that is right in front of our eyes. Mostly, we can only guess what it might be like to see the world as other animals do. The closest approximation of how astonishing it could be might be likened to a colourblind person putting on EnChroma glasses and seeing colour for the very first time. Quite often, their jaw drops as they look around in utter amazement at colourful flowers and lush green trees. The experience can be overwhelming, and often they burst into tears.

---

7 Mysteriously, the animal with the greatest number of opsin genes is Daphnia pulex, the water flea, whose genome codes a whopping 46 opsin genes.

Another glimpse of the world we're blind to comes to us from people with a rare condition called tetrachromacy. Tetrachromats, as they are known, have a sort of hyper-colour vision, the ability to see a richer and more vibrant world than the average person can see. That's because they are born with four different cone cells for colour vision whereas most of us have three. This fourth receptor allows them to perceive ninety-nine million more shades and hues than the average eye perceives. The genetic mutation is found in about 12 percent of women, but only a small sub-set of this group have true tetrachromacy.

So what does the world look like when it's one hundred times more colourful? Tetrachromat Concetta Antico has described it as "seeing colors in other colors." Compared to her, we see the world almost like the colourblind do. What is a grey pebble pathway to our eyes lights up in hers with a rainbow of different hues. As she describes it, "The little stones jump out at me with oranges, yellows, greens, blues and pinks." Beyond an appreciation of beauty, this form of sight offers practical utility. When asked what she can see that others can't, she has said, "I can tell if someone is sick just by looking at them. Their skin gets gray, it gets yellow, and there's some green. I can tell when my daughter is sick because she will be all washed out and greenish-yellow or maybe whitish-lilac." But we'll never really know what she means, since she's just using colour to describe another colour. And colour for her means something different from what it means for us.

Individuals with aphakia can also see beyond the ordinary. Like bees and eagles, they have the ability to see in ultraviolet. The condition is often a result of eye surgery, although at times it can be caused by a congenital anomaly. The term *aphakia* comes from the Latin, meaning "no lenses." The reason most of us can't see in ultraviolet is because the human lens naturally blocks out UV light. But patients who have undergone cataract surgery and had the lens removed can sometimes see into this range of the spectrum. Perhaps the most famous aphakic was Claude Monet. In 1923, at the age of eighty-two, the impressionist artist had his left lens removed during cataract surgery. When Monet began painting water lilies again, they were no longer white but had tints of deep purples and whitish blues. But again, we're not seeing what he saw. He painted white lilies mauve and lilac, but mauve and lilac looked different to him than they do to us. Still, whatever colour the lilies were in his eyes, they were almost certainly not white.

By the Second World War, military intelligence had become aware of this real-world superpower and used aphakic patients as coastline lookouts. At the time, German U-boats used UV lamps to send covert signals to their onshore agents. The aphakics were enlisted to send in alerts when they saw the lights, which were invisible to everyone else. This should give us a pretty strong sense of the size of our perceptual blind spot. There can be enemies right off the coast, invisible to us but obvious to those who can see.

But not having the ability to see something doesn't mean you can't look. As an avid surfer, Mike Sturdivant had spent over thirty years in the water off the Gulf Coast of the United States. But in July 2010, something strange started to happen: he began coughing up blood. And Mike was not alone. Along the Florida beaches, people were starting to complain of shortness of breath, burns on their skin, and blurry vision. It was clear to Sturdivant that there was something in the water.

One night, he decided to take his UV light, which he used on his boat to check for engine leaks, to scour the beach and see what he could find. What he saw was confounding: from "the dune line to the water line," the entire beach was glowing a bright orange.

Over two hundred kilometres away, cleanup operations were under way for the largest marine oil spill in US history. Over four million barrels of oil had spilled into the Gulf of Mexico from the *Deepwater Horizon* drilling rig, and an additional 1.8 million gallons of Corexit brand dispersant were poured into the water in an effort to make the oil degrade more quickly. Scientists would later discover that the combination of the oil and dispersant made the water fifty-two times more toxic.

Illuminated under a 370-nanometre UV light, the toxic mixture fluoresced. A year after the spill, Sturdivant partnered up with James Kirby, a coastal geologist at the University of South Florida, to begin a formal investigation. Over two years, the duo sent seventy-one samples to a lab for testing. The results were what they suspected.

Under the National Contingency Plan for responding to oil and hazardous waste spills, a beach is considered clean if it contains less than 1 percent of oil visible in a one-square-metre sampling area. But dispersant doesn't remove oil; it disperses it. According to Sturdivant, that's the problem: "The entire [cleanup] operation has been geared around making things invisible. And that's why they're using the dispersant. It's not because it will help speed up the degradation of the oil. It's because it makes it invisible." That is, it makes it invisible to humans. Some animals, of course, can still see it perfectly well.

But even if we can't see it, we can still feel it. Years later, Gulf residents are still complaining of strange symptoms: skin rashes, migraines, nausea, seizures, rashes, bloody diarrhea, pneumonia, muscle cramps, severe mental fuzziness, and even blackouts. To the naked eye, however, the Florida beaches look picture perfect.

IN THE ANIMAL WORLD, sight comes in other forms as well. Beyond the ability to see heat, see in ultraviolet, and see Earth's magnetic fields, there's also the ability to see by means of using sound, or echolocation. Bats and toothed whales evolved this ability independently. Either in the air or underwater, by emitting a series of rapid-fire buzzing or clicking sounds and listening for the echoes, animals are able to determine the shape, location, and movement of objects around them.[8] Bats have an acoustic visual field of about two to ten metres away and can "see" as close as four to thirteen millimetres, which is important for hunting small insects. Common bottlenose dolphins have a biosonar range of about 110 metres, while sperm whales, who hunt squid in the depths of the ocean, have the largest field of view, able to spot prey at 500 metres.

So how do we know that biosonar allows animals to "see"? The first study to look into bats' ability to fly in total darkness was done in the eighteenth century by Lazzaro Spallanzani. Spallanzani was determined to figure out which sense the bats[9] were using, so he isolated each one—vision, touch, smell, taste, and hearing—and eliminated them one by one.

It is, of course, a myth that bats are blind, but to ensure it wasn't their sight that allowed them to avoid obstacles in the dark, Spallanzani blinded them, first by covering their eyes with a hood and then, more gruesomely, by removing their eyes. In his notes, he wrote, "Thus with a pair of scissors I removed completely the eye-balls in a bat. . . . Thrown into the air the animal flew quickly, following the different subterranean pathways from one end to the other with the speed and sureness of an uninjured bat. More than once the animal landed on the walls

8 More specifically, by listening for the returning signal strength, the direction and time for the echoes to bounce back from an object, the brain is able to triangulate and form a shape-image of the object.

9 In 1938, a Harvard undergraduate student, Donald Griffin, used a sonic recorder to hear the sounds bats were making that were above the frequency range of human hearing. This was the first proof that bats use echolocation.

and at the roof . . . and finally it landed in a hole in the ceiling two inches wide, hiding itself there immediately. My astonishment at this bat which absolutely could see although deprived of its eyes is inexpressible."

Studies with dolphins—thankfully with their eyes intact—have also led to profound insights about their abilities. Controlled studies on captive dolphins have found that they can recognize distinct shapes by using biosonar alone. Researchers at Kewalo Basin Marine Mammal Laboratory in Hawaii put a dolphin named Elele to the test by placing objects of various shapes inside a box. The box was made of a thin black Plexiglas that was opaque to the eye but could be penetrated by sound. Elele was shown three objects being held in the air by the trainer and asked to identify which one matched the object in the box by pointing with her rostrum to the matching object. She performed exceptionally. Able to switch senses with ease, Elele could "see" what object was inside the box whether she used vision to match the echolocated object or echolocation to match the visible object.

Anecdotally, people have long observed that dolphins are particularly curious about pregnant women and other pregnant dolphins, swimming up and making a buzzing sound near the belly of the expectant mother. While not confirmed, it wouldn't be surprising if dolphins can "see" right through flesh and into our bodies, as the ultrasound that dolphins use for echolocation is similar to the ultrasound that medical practitioners use to image fetuses.

And what does a "click image" look like? It's impossible to know. But if anyone could give us a clue, it would be Daniel Kish, the real-life Batman. Blind since he was a baby, he began clicking with his tongue and listening for the echoes as a way of creating a mental picture of the world. It wasn't until he was eleven years old, when a friend asked him if he was using echolocation, that Kish realized he was doing what bats did in order to "see."

While humans don't have a bat's fine-tuned abilities to detect small, quick movements, Kish has developed some remarkable sensitivities. He can hear buildings, and in a click "see" whether they are ornamented or plain and featureless. In an auditorium, he can hear the exits, and he usually knows where they are before a sighted person spots them. He rides his bicycle around the city using only his ability to echolocate. Researchers studying Kish's brain activity using MRI have found that the area of the brain being activated when he echolocates is the region typically devoted to vision. What that tells us is that his brain registers sound as sight. He not only hears sound, he "sees" it.

Few of us will ever know what it's like to see like Daniel Kish. But knowing what we don't know tells us something. Kish inhabits a sensory world as unknowable to us as a bat's or a whale's. It is the same world as ours, and at the same time it is completely alien to us. Yet Kish is unquestionably human. Which tells us that while the way our fellow animals see the world may seem exotic or alien, there is no sense in which we could reasonably believe it's inferior.

THE SENSE OF SIGHT is not only singular however, it is communal in that it allows us to mimic and learn from others. Children watch adults intently to copy what they do, and the same is often true of animals. This is why, using sight, a bee with a brain the size of a sesame seed can be taught to do something it would never naturally do in the wild: it can learn to play soccer.

Using a plastic bee attached to the end of a stick, researchers at Queen Mary University of London first showed the trainee bees what to do. The bumblebees watched as the fake bee pushed a small ball into a circle. This was the "goal." When the ball was manoeuvred inside the lines, the insects were given a sugar-water treat. After three observation trials, the trainee bees were then placed in the miniature stadium. Just by watching the fake bees, they were able to mimic the unnatural task, scoring a goal themselves 99 percent of the time.[10]

Sight is also key to visual memory. And the animal with the most jaw-dropping visual memory known to science, is Ayumu, a chimpanzee who lives at the Primate Research Institute at Kyoto University. What makes Ayumu special is that he has an eidetic, or photographic, memory. In less than the blink of an eye, his mind can absorb a full scene, and going head to head with a human on a visual memory task, he will win every single time.

On the surface the task is straightforward. It looks something like this: picture the numbers 1 to 9 placed in a jumbled order in random places on a screen. All nine numbers flash up at the same time and are held there for just over half a second. After the numbers flash off, blank white blocks light up where the numbers were and remain in their place. The task is to tap, as fast as possible, in correct sequence, the white blocks where the numbers from 1 to 9 previously appeared.

To watch Ayumu do it is stupefying. Even after looking at the screen for several seconds, a human finds it hard to process the placement of just a few numbers, let alone all nine. In a test against British memory champion Ben Pridmore, who can memorize the order of a shuffled pack of card in less than thirty seconds, Ayumu was the clear winner. Pridmore had an accuracy of only 33 percent, while Ayumu was correct 90 percent of the time. But for Ayumu, even half a second is a considerable amount of time: he's been able to succeed in this memory task after seeing the images for just 210 milliseconds.

---

10 Untrained bees only "scored" by chance: 30 percent of the time.

So how is he able to do this? We know that, in general, chimpanzees are better than we are at something called subitizing, which is the ability to see and instantly know the number of objects in view, similar to how you can see the number 4 on dice without having to count the dots. While humans can subitize up to four or five random numbers, chimpanzees can subitize upwards of six. Ayumu is particularly good at this, even for a chimpanzee. He outperforms both humans *and* chimpanzees, suggesting that his visual memory is outstanding.

Animals do not just process what they see like automatons, of course. They are active agents. And like humans, they communicate what they see in the world around them. And while almost all communication occurs within species, some scientists have ventured across boundaries and are learning to see through animal eyes by teaching them to communicate with us.

The most famous communicator from the avian world was an African grey parrot named Alex. Selected at random from a pet store, he was reared by researcher Irene Pepperberg and became so well known for his abilities that he revolutionized our ideas about animal intelligence.

Although Alex had a brain the size of a walnut, he had the cognitive abilities of a five- to six-year-old child. Pepperberg trained Alex to answer questions about what he saw. He was shown objects and taught the words that described them. While parrots don't have a larynx with vocal chords, they do have a syrinx, allowing them to mimic the sounds of human speech. Alex could identify different shapes and colours and count to eight; he knew the difference between "same" and "different," and "bigger" and "smaller"; and he could communicate with more than a hundred words.

Alex would also come up with labels for novel things he encountered. In the process of being taught fruit names, for example, he surprised the researchers. He already knew the words "banana," "grape," and "cherry," because he learned the names of the fruits that he was fed. But when he saw an apple for the first time, the bird had his own word in mind. He insisted on calling it a "banerry." Why? Well, it's possible that Alex saw it was red on the outside and yellow in the middle, or he tasted it and decided to use a combination of the two fruits he already knew: a banana and cherry. Either way, "banerry" it was, because from then on he refused to call it an apple.

Alex is also the only animal known to have asked a question about himself. In December 1980, he caught sight of his reflection in a bathroom mirror. Turning towards it, the parrot asked his handler, Kathy Davidson, "What's that?" Kathy replied that was him, Alex, and that he was a parrot. After studying himself for a while longer, he asked, "What colour?" Kathy said, "Grey. You're a grey parrot, Alex." After a back and forth that went on a few more times, Alex, it seems, finally understood. According to Pepperberg, that was how Alex learned the colour grey.

This ability to describe the visual world is not unique to parrots. Other animals have learned to communicate with us as well. Perhaps most famous was a gorilla named Koko. Using a modified version of American Sign Language (ASL), Koko had an extensive vocabulary: she was able to sign over one thousand words and understood over two thousand words in English. Like Alex, Koko was been known to come up with words for new things that entered her environment. The first time she saw a zebra, for example, she described it as a "white tiger"; a Pinocchio doll was signed as "elephant baby"; and the first time she saw a ring she called it a "finger bracelet."

It should be noted that scientists still rigorously debate whether animals like Koko and Alex truly had the ability to communicate. That's because science demands strict objectivity in verifying results, but language is subjective and, as we all know, often ambiguous. When it comes to studying animal intelligence, scientists often rely on a principle called Morgan's Canon. Essentially, it states that one should not ascribe higher psychological processes to an animal if the same behaviour can be attributed to something simpler, like an error.

Eugene Linden, the author of *Apes, Men, and Language*, describes the same situation with Washoe, a chimpanzee, and the first ape to learn ASL: "About 50 years ago, on a pond in Oklahoma, Washoe saw a swan and made the signs for 'water' and 'bird.' Was she simply noting a bird and water, or was she combining two of the signs she knew to describe an animal for which she had no specific word? The debate continued for decades and was unresolved when she died."

There may be one way to settle the debate, however, and that's by controlling for the environment that animals are being questioned about. In Norway, scientists came up with a clever way to do this by training horses to communicate with symbols. The task was simple: the horses were trained to use their muzzles to point at a board, indicating whether they wanted to wear a blanket. A vertical bar meant "take the blanket off," a horizontal bar meant "put the blanket on," and a blank symbol indicated "no change" as their preference.

After just two weeks of training the animals for fifteen minutes a day, they were able to use the signs to communicate. The horses were not simply discriminating between the visual cues; they were making their decisions based on the outside weather. On warm days, when the temperature was 20°C to 23°C, all ten of the horses given blankets requested that the blankets be taken off. The horses that did not have blankets on, signalled that they wanted their state unchanged. On rainy, cold days, when the weather was 5°C to 9°C, all ten of the horses with a blanket on signalled that they wanted their state unchanged, whereas ten of the twelve horses not wearing blankets requested to have one put on.

That twenty of the twenty-two horses wanted a blanket on cold days suggested to the researchers that the horses did indeed understand the visual symbols and were making requests. As for the two holdouts, on colder testing days, where the temperature ranged from −12°C to 1°C, the outliers gave in and joined the rest.

While these "talking horses" are impressive, the world's best animal communicator, at least when it comes to using visual symbols, is Kanzi, a male bonobo, who lives at the Great Ape Trust in Des Moines, Iowa. Kanzi has a five-hundred-word vocabulary in the form of touch screen symbols called lexigrams, he understands over three thousand English words, and is said to comprehend complete sentences and instructions.

Wearing a welder's mask so that her facial cues or eye movements didn't give anything away, Kanzi's trainer Sue Savage-Rumbaugh conducted tests by coming up with novel sentences and bizarre requests to see exactly what Kanzi understood. Asked to put salt on a ball, Kanzi promptly took the salt shaker and did as requested. When Savage-Rumbaugh asked Kanzi to put pine needles in the refrigerator, Kanzi again had no problem performing the task. Asked to carry the TV outdoors, Kanzi got up, looked around, spotted the television set, and immediately carried it outside.

To understand just how astonishing this is, pause to consider what Kanzi might have been thinking. Many humans have difficulty picking up a foreign language. These animals are dealing not only with a foreign language but the requests of a foreign species. If Kanzi is smart enough to learn what many of us cannot, it's not implausible to imagine that he may have wondered why Savage-Rumbaugh would want to refrigerate pine needles in the first place. What might he be making of the human in front of him before getting up and, once again, complying?

At this point, it's fair to ask: if bonobos and chimpanzees can do it, can we? Seals and dolphins understand our hand signals, dogs and elephants understand our vocal commands, orangutans can even use iPads to communicate with us. But what do *we* understand about other animals' languages? What are they seeing and describing in their own tongues? As science journalist Rachel Nuwer has observed, in trying to "force apes to learn our language, we may have blinded ourselves to theirs." To find out, one man has spent much of his academic career taking the time to flip the script, and in doing so he has opened up a whole new way to examine how animals communicate on their own terms. His name is Con Slobodchikoff, and he has been called a modern day Doctor Dolittle.

Slobodchikoff, an emeritus professor of biology at Northern Arizona University, works with Gunnison's prairie dogs, highly vocal little animals that look like North America's version of meerkats.[11] Peeping their heads up from their burrows, the prairie dogs are often on high alert for predators. Noting that they give off alarm calls, Slobodchikoff began recording the sounds they made when they saw different predators approach. To our human ears, the calls for the most part are the same: short barks in quick succession that sound almost like they are coming from a squeaky toy. But a computer analysis revealed something else: each call was unique. And by visualizing the sound waves of the calls in a sonogram, Slobodchikoff could see that the barks for the different predators were clear and distinct.

Barks for "human," "hawk," "coyote," and "dog" all have signature acoustic sonograms with differing wavelengths and amplitudes. And despite some of the predators looking similar, the prairie dogs never bark "dog" when they see a coyote, or vice versa. For Slobodchikoff and his research team, the sonograms were a way to decode the rodents' communication, like a prairie dog Rosetta stone.

---

11 Meerkats belong to the mongoose family while prairie dogs are rodents. So while they look similar, they are very different.

So how can we be sure that the calls mean what we think they do? Slobodchikoff and his research team not only recorded the prairie dog alarm calls but also video recorded their escape responses. Upon seeing a hawk, the animals would look up, make a quick one-syllable bark, and scurry down to their burrows. An audio playback of the prairie dogs' "hawk" bark would elicit the same response: they would look up and search in the sky, then run towards their burrows. When the sound of a dog was played, however, the prairie dogs would stand on alert but not run away.

Robert Seyfarth has done similar work with vervet monkeys. The primates have different alarm calls, or "words," for hawks, snakes, and leopards. The monkeys respond to the "leopard" call by running up a tree, but on hearing the alarm call for "hawk," they looked up in the air and quickly sought safer ground by hiding in a bush. They avoid climbing trees during "hawk" calls, presumably because in a tree the raptors catch them. The calls are clearly meaningful among the troupe. When a vervet monkey makes an alarm call for "snake," the primates all stand on their hind legs and begin scouring for signs of the predator lurking in the grass.

Slobodchikoff took this observation one step further. He wanted to see how prairie dogs reacted to something abstract that they'd never seen before, so he built plywood cut-outs of circles, squares, and triangles. Then he and his team strung a cord from a tree to their observation tower, hung the shapes on them about one metre off the ground, and pulled them through the colony like clothes on a laundry line. The prairie dogs responded to the new "threats" coming through with different barks. Incredibly, the animals had distinctive calls for "circle" and "triangle," even though for the colony the shapes were entirely new.[12]

---

12 The prairie dogs seemed unable to tell the difference between a square and a circle.

Slobodchikoff had also observed that the prairie dog calls appeared nuanced. He wondered if there was in fact more information in each call, if the call for "dog" was the same for all dogs, or if the calls would be different depending on the breed. So he sent four different breeds through the colony: a golden retriever, a husky, a Dalmatian, and a cocker spaniel. When he scrutinized the calls, he found that the prairie dogs' barks were indeed more than simple alarms indicating "dog." He had a hunch that they might be descriptors.

With people acting as the colony intruders, Slobodchikoff began to record striking differences in the calls. The prairie dogs had different calls for tall humans and short humans. If the people were shaped differently, the barks were reflective of whether the person was fat or thin. And finally another incredible distinction emerged: the rodents had specific barks depending on the colour of the clothes people were wearing.

By controlling the variables, Slobodchikoff could find out what was happening. He had his lab assistants walk singly through the colony, changing one variable: the colour of their T-shirts. The same people would walk through the colony wearing shirts that were either blue, green, or yellow. The results were nothing short of mind-blowing: the prairie dogs barks were describing the intruders.

Slobodchikoff had deciphered what the animals were saying to each other *about* us. When an assistant wore blue, the prairie dogs barked, "Tall, thin, human, blue," and when just the color of the shirt changed, the prairie dogs said, "Tall, thin, human, green."

Our bubble is the belief of human exceptionalism: that we are the only species aware enough to feel, think, and speak. As Slobodchikoff's studies show, prairie dogs can accurately describe the world around them, not because they are trained to use labels, but because they are naturally communicating what they see.

IT'S IRONIC THAT WE SAY someone is as "blind as a bat," because bats in fact have two ways to see. The man who discovered this, and who in 1944 coined the term "echolocation," was zoologist Donald Griffin. He spent the first half of his academic career studying the remarkable traits of this sonic vision and the latter part focusing on a particular form of human blindness: the belief that we are Earth's only aware and sentient beings. This blind spot has been strong in the sciences, particularly among animal behaviourists, who until recently lobbied against evidence of animal consciousness, calling studies that supported such evidence groundless and "unscientific."

Like many thinkers of the past who have challenged the status quo, Griffin met with a flood of criticism in response to his initial work in this area. His 1976 book *The Question of Animal Awareness* was later called "*The Satanic Verses* of animal cognition" by one of his critics. There were those within Griffin's field who lamented that a once-great scientist had fallen and that this new quackery about animal awareness was likely the mark of "premature senility." We would do well to remember, of course, that the greatest scientists have always questioned humanity's central role. The same central tenet of this idea—questioning that the universe revolves around us—is what got Copernicus banned and Galileo jailed.

Human exceptionalism is notoriously persistent, however. We refer to animals as though they were objects. An animal is an "it." The idea that animals are subhuman, that they lack awareness and intelligence, that they are *inferior*, has led us to treat them as though they are not only property but biological machines. In the early years of laboratory testing on animals, the logic was that animals were not "feeling" but simply reacting. That when "a dog screams when its body is hurt, its vocalization is not the expression of pain but merely the result of a purely physiological process, rather like the ringing of a clock." As if our own pain weren't physiological.

For the primatologist Frans de Waal, this type of thinking is a form of neo-creationism, like a decapitated theory of evolution. As he writes, it "accepts evolution but only half of it . . . It views our mind as so original that there is no point comparing it to other minds except to confirm its exceptional status." It's as if evolution stops at the head, and yet, when it comes to bodies, we feel more confident about what we put in or on our bodies after we have tested it on animals. Indeed, we test drugs on animals before human trials precisely because we believe that the effects can be extrapolated, *because* of our similarities.

And yet it's important we also respect that we do have differences and that just as it's impossible to be in another human being's head and know how they see the world, it is impossible to know truly how a bat, or a chimp, or a dung beetle envisions the same world. The American philosopher Thomas Nagel famously noted this in his essay "What Is It Like to Be a Bat?":

> Even without the benefit of philosophical reflection, anyone who has spent some time in an enclosed space with an excited bat knows what it is to encounter a fundamentally alien form of life. . . . Bat sonar, though clearly a form of perception, is not similar in its operation to any sense that we possess, and there is no reason to suppose that it is subjectively like anything we can experience or imagine. This appears to create difficulties for the notion of what it is like to be a bat. . . . I want to know what it is like for a bat to be a bat. Yet if I try to imagine this, I am restricted to the resources of my own mind, and those resources are inadequate to the task.

Our minds are alien to other life forms on Earth, just as their minds are alien to our own. And though we may think that animals like our pets know when we are happy or comfort us when we are sad, we are assuming they can make a mental leap that we ourselves are unwilling to make in the opposite direction.

While we may never know what another animal is feeling or thinking, it is no longer a scientific stretch to say that they do feel and think. We have seen grand revolutions in scientific thinking, but a certain dogmatism still remains when it comes to animal intelligence. Thankfully, the rigid notions of our speciesism are slowly fading away. On July 7, 2012, the Cambridge Declaration on Consciousness was signed by a prominent international group of cognitive neuroscientists, neuropharmacologists, neurophysiologists, neuroanatomists, and computational neuroscientists. Together, they declared that "convergent evidence indicates that non-human animals have the neuroanatomical, neurochemical, and neurophysiological substrates of conscious states along with the capacity to exhibit intentional behaviors.... The weight of evidence indicates that humans are not unique in possessing the neurological substrates that generate consciousness. Non-human animals, including all mammals and birds, and many other creatures, including octopuses, also possess these neurological substrates."

The old saying is that "the eyes are windows to the soul." In science, the existence of a soul may be untestable and unverifiable, but the existence of a consciousness is not. Our own eyes are a window into only one way of seeing the world, a sliver of consciousness among millions of other unimaginable ways of perceiving it.

We cannot trust our senses when it comes to perceiving the bigger picture of reality. Indeed, when it comes to what surrounds us, we have already revealed three big blind spots. Our naked eyes and common sense would have us believe that we are the centre of the universe, isolated and separate from the world around us, and superior to all other creatures. But with the corrective lens of science all three of these assumptions can be overturned.

We have mastered sight in another way, however. We are a singular species, with cameras and high-tech eyes everywhere.

We have the technological lenses to see into the vast distances of outer space, to see the tiniest microscopic organisms, to see right through the human body, and to see the very atoms that make up the material world. But there is one fundamental thing that we do not see. When it comes to how our species survives, we are utterly blind.

PART TWO

SOCIETAL BLIND SPOTS

WHAT SUSTAINS US

4

RECIPE FOR DISASTER

Think, occasionally, of the suffering of which
you spare yourself the sight.

—ALBERT SCHWEITZER

THE BODY ON THE AUTOPSY TABLE was unrecognizable. What was once a living, breathing being had been radically transformed. It was up to professor of medicine and pediatrics Richard deShazo, along with two pathologist colleagues, to do the examination. Published in the *American Journal of Medicine*, the Mississippi study was a first: they were about to slice open and dissect a chicken nugget, for science.

Fixed in formalin, the fast food was carefully sectioned, stained, and placed under a microscope. Troubled by Mississippi's growing obesity epidemic—Jackson has the highest obesity rate in America, with over a third of the population severely overweight—the team wanted to know more about food in the urban centre and exactly what it was that people were eating.

What the researchers discovered "floored" and "astounded" them. Striated muscle, or chicken meat, "was not the predominate component" of the nuggets at all. The nuggets were mostly fat, bone, epithelium (the cells that line the organs and skin), nerve, and connective tissue. The remaining 40 percent was skeletal muscle.

The chicken, or more likely chickens, in each nugget had been transformed into a batter-like paste. Known in the industry as "mechanically separated poultry," this was tissue that has been forced under high pressure to separate it from the bone. As deShazo explained in an interview, "You can actually vibrate that stuff off, and you get these chicken leftovers, and you can put it together, mix it up with other substances, and come out with a goo that you can fry and call a chicken nugget. It's a combination of chicken, carbohydrates, and fats, and other substances that make it glue together. It's almost like superglue that we're eating."

Sometimes we do actually eat glue, a delicious little confection that goes by the appetizing name transglutaminase, or TG. Humans have this enzyme—when you scrape your knee, it's what allows the blood to clot—though the commercial version is either synthesized from bacteria or made from the blood plasma of cows or pigs. And just as it can mend your knee, it can bind the proteins in scraps of meat together so that separate bits can be shaped into a solid piece that looks like a fancy cut. In a similar vein to how Dr. Frankenstein's monster was stitched together from different body parts, "Frankenmeat" is assembled from leftover body parts, sometimes of separate animals.[1] It works so well that even trained butchers can have a hard time spotting a loin that has been made from separate scraps. In the food industry, the most common "restructured meat" is filet mignon. At cheaper banquet halls and hotels that serve in bulk, this insider trick cuts the costs on an expensive cut of beef.

When it comes to meat, things are almost never what they seem. All meat is dead, of course, but some is a little more dead than most. In 2015, Chinese authorities cracked down on a fourteen-province-wide "zombie meat" smuggling ring. Customs officials confiscated one hundred thousand metric tons of frozen pork, chicken, and beef that dated back to the 1970s and 1980s and that was being sold off to local food stalls and restaurants. According to the *Hong Kong Free Press*, the forty-year-old meat had been "pumped full of chemical additives to keep them looking fresh." In Chongqing, the epicenter of the smuggling ring, a cover for the spoilage was the fact that the region is renowned for spicy cuisine. If the meat had any suspect taste it remained somewhat well-masked, though a more serious issue was that the old meat could also have been diseased, as it originated in areas potentially affected by bird flu, foot and mouth disease, and mad cow disease. Frozen-food smuggling is a high-profit trade. The haul from the Chinese bust was worth a total of ¥3 billion (US$430 million), leading inspectors to believe that it wouldn't be their last.

---

1 Examining thirty cookbooks by prominent chefs, Ike Sharpless published a study entitled, "Making the Animals on the Plate Visible: Anglophone Celebrity Chef Cookbooks Ranked by Sentient Animal Deaths." Ranked by order, it was found that *Batali's Molto Gusto: Easy Italian Cooking* was the worst offender, with 5.25 average deaths per recipe and 620 total animal deaths.

While zombie meat is well past its expiry date, what we consider "fresh" is still relative. The tuna glistening on ice under halogen lights in your supermarket looks fresh enough, but it could have been caught weeks or months ago and shipped halfway around the world and back after being frozen and thawed a couple of times. Because the bright red of tuna naturally fades into an unappetizing brown, imported fish is often gassed with carbon monoxide to prevent the flesh from discolouring in transit. While the process itself is harmless, it can lead to health risks, since simulating freshness can potentially mask spoiled fish. It also deceives the consumer, who is unable to tell whether the fish they are buying is a month old or has just been caught.

Carbon monoxide won't make farmed salmon look more delicious. Wild salmon are pink-fleshed because they eat wild food: krill and microalgae. Farmed salmon are fed soy- and corn-based diets. As a result, their flesh is not pink but grey. But would you buy grey salmon? Merchandising experts suspect you would not, so fish farmers use what's called the SalmoFan, a Pantone-like fan of colours, much like the paint chips used in interior design, so that farmers can create an appropriately pink salmon. Launched by the Royal DSM company in 1989, it is "the industry's color reference standard for the visual judging and comparison of degrees of pigmentation in salmon flesh perceived by the human eye." The process is known as "colour finishing," and the colours you can select for your salmon come in fifteen different shades, from a soft pink to a rich red-orange. Today, 70 percent of salmon in the global market is farmed, and all of it is artificially coloured with canthaxanthin and astaxanthin, which are synthetic carotenoids made from petrochemicals.

For eggs, the same company sells a YolkFan, which provides, as its name implies, a palette of sixteen colours to measure egg yolks. In Asia, customers prefer a paler yolk, while in countries like New Zealand, shoppers prefer a deep orange. To cater to different geographic preferences, egg farmers who want the perfect "golden hue" can add Carophyll red and Carophyll yellow to the feed for caged birds that do not forage outside. Most people assume they can tell the difference between eggs from a pasture-raised hen and a factory-farm raised hen, based on the richness of yolk colour. But with feed additives, we can be duped, and colour alone is no longer an indicator of a healthy egg. In fact, colour is just one more facet of marketing.

Fake freshness in the food industry can be traced to its beginnings in the 1950s and '60s, when food scientists began coating meat with antibiotics. As Maryn McKenna writes in her book *Big Chicken*, "Hundreds of scientists experimented with coating meats and fish in antibiotic solutions, misting the drugs onto fruits and vegetables and mixing them into milk." The process was called "acronization," and became a favoured method for preserving chicken. After butchering, the birds were soaked in an antibiotic solution. By preventing bacteria from spoiling the flesh, processors were able to increase the chicken's shelf life, and the time it could be kept for sale.

The method came to a rather ignominious end, however, when slaughterhouse workers began getting staph infections, with boils and lesions searing their arms and hands. It was not the antibiotics themselves that caused the infections but rather strains of bacteria that had become resistant to acronization. The process was terminated shortly thereafter. Today, birds are no longer soaked in antibiotics, but in the United States they are soaked in something else: chlorine. While it sounds repulsive, it is actually safe to eat chorine-washed chicken, as long as the concentration of chlorine remains low, in a solution of twenty to fifty parts per million. The method kills off food-borne pathogens like campylobacter and salmonella to ensure that they don't survive and spread after slaughter. But in essence, chlorine is a chemical blind spot. It prevents us from seeing what we otherwise might not ignore.

Across the pond, the United Kingdom banned chlorine-washed chicken for reasons that have less to do with health risks and more to do with systemic sanitation. Because there are fewer welfare protections for the birds in the United States, more of them can be crammed into cages or grow-out houses, which often results in sicker birds and a greater spread of fecal matter and disease. The chlorine rinse is the safeguard that ensures that the bacteria are washed off before the birds go to market. In the United Kingdom and European Union, however, the logic is reversed. The minimum allowable amounts of space, light, and ventilation required for hens are greater in Europe than in the United States. The US minimum space requirement per bird is half a square foot (465 cm^2). The UK minimum is double that. Either way, it's not a lot of room for birds, particularly broilers, who are engineered to be relatively large, six-pound animals.[2]

All this is to say, when it comes to what we eat, our eyes often deceive us.

FOR YEARS, PEOPLE HAVE TRIED to justify the five-second rule. The "rule" suggests that if food falls on the ground, you have five seconds to pick it up before it becomes bacterially

---

2 The extra space in Europe however does mean the birds are slightly freer to move. With birds that are not as sick and soiled, there is less need for chemical decontamination. The same is true for the difference in eggs. You may have noticed that eggs in US supermarkets are always sold refrigerated, while in Europe they are stocked on shelves at room temperature. That's because the egg-laying conditions in the US eggs are dirtier, so they have to be sprayed with a chemical sanitizer. After being rinsed in hot water, the eggs must be kept refrigerated afterwards.

contaminated. Of course, there is no science to back this up. Instead, we come up with our own justifications like, "Don't worry. It's just a slice of cheese. You can wipe it off quickly," or "It's a jelly bean, not a gummy bear. See? Nothing stuck to it." The science however, is definitive: drop your food on the ground and almost instantaneously it will have bacteria on it. So why does the myth persist? Simply, because we want it to. We can't see the bacteria and it doesn't appear to cause any harm, so most people (79 percent, according to one survey) will pick up and eat food that's been dropped on the floor. Now, a dirty jelly bean is one thing, but when it comes to the dirty truth of our food system, are we able to confront the facts or do we do the same thing and look away because we want to?

With food, there are things we'd rather not know. And here's the rub: we *know* we'd rather not know. Scientists have found that our brains will shut down information that doesn't make us feel good, or that causes us stress, which is one reason why we tune out suffering. But as Margaret Heffernan writes in her book *Willful Blindness*, "Not knowing, that's fine. Ignorance is easy. Knowing can be hard but at least it is real, it is the truth. The worst is when you don't want to know because then it must be something very bad. Otherwise you wouldn't have so much difficulty knowing."

If we were the least bit curious to know where our food comes from, it wouldn't be hard to get the facts. The gothic horrors of the meat-packing industry have been well known since Upton Sinclair published *The Jungle* more than one hundred years ago. Though you might be a little less likely to find a rat in a tin of corned beef today, the sheer scale of the slaughter has grown enormously, and mechanization over the past century has arguably only made slaughterhouses and industrial-scale farms more shocking. As James Pearce notes in his essay "A Brave New Jungle," "Perhaps the most insightful way to illustrate the intensification of animal-intensive agricultural production over the course of the twentieth (and into the twenty-first) century is with a simple statistic: the poultry industry today slaughters more birds in one day than the entire industry did in the year of 1930."

And while those who profit from the carnage are content to keep the facts and statistics hidden, unpleasant truths are the easiest things in the world to hide. If someone doesn't want to know something, they're not going to know it.

Disgust is another powerful inhibitor. Disgustologists, as the scientists who study the subject like to call themselves, have found that the emotion of disgust is universal, and it does have benefits. That we recoil and grimace when we see sores or lesions on putrid flesh, for instance, is an evolutionary advantage. Disgust keeps us away from pathogens. It protects us from disease.

But many things we might potentially be disgusted by are no longer in view. In particular, when it comes to cheap meat in the food industry, we are in the dark about important facts about our food. Facts like, the animals we eat are routinely fed garbage and other animals' feces.[3] Facts like, most bacon comes from pigs that were put in a gas chamber. And facts like, there are steaks in supermarket meat cases that came from a steer that was skinned alive.

Perhaps you'd rather not know that. Some facts definitely make it more difficult to eat, or at least to shop. It is certainly not suitable conversation for dinner. Reading off the ingredients in, say, artificial coffee creamer is unappetizing enough, but it's a different

---

3 "The word "garbage" isn't proverbial. Mixed in with the grain can be an assortment of trash, including ground glass from light bulbs, used syringes and the crushed testicles of their young. Very little on a factory farm is ever discarded."

—Solotaroff, Paul. "In the Belly of the Beast" *Rolling Stone*. (2013, December 10).

matter entirely to contemplate the provenance of a cutlet. And while we are content to know little to nothing about everyday ingredients—dipotassium phosphate, mono- and diglycerides, silicon dioxide, sodium stearoyl lactylate, soy lecithin, and artificial flavours—in the case of food that was once alive, knowing little to nothing is a different kind of opacity and at least partly a matter of conscience. Not knowing is a way to keep our consciences clear.

Think back to the previous chapter and the animals whose inner lives and sensory experiences are as rich as ours. We are animals too, after all. And animals tend to have an innate regard for other animals. The esteemed American biologist E.O. Wilson called this "biophilia," or "the urge to affiliate with other forms of life." Many humans feel a sense of reverence or connection when in the presence of nature, perhaps because we *are* a part of nature. It is all but impossible to resist the desire to nurture and protect a puppy or kitten; few can approach a horse without wanting to put a hand on its neck. A sense of kinship with fellow animals is not the subject of rigorous scientific study, but the evidence for it is impossible to ignore. We love animals.

Deep down, our similarities are difficult to deny. A cow's experience, or a chicken's, or even a bat's, as we've already seen, is certainly different from our own. We don't know what it's like to be a cow or a chicken or a bat, but it's *like* something. When we think about what it's like to be an animal, we are in the same position a robot with artificial intelligence (or a Martian) would be in when processing the question of what it's like to be human. That is, the fact that our behaviours can be dispassionately described doesn't mean it should be assumed we are incapable of rich experiences. And we have no grounds to make that assumption about other animals. As Thomas Nagel points out, "To deny the reality . . . of what we can never describe or understand is the crudest form of cognitive dissonance." We can't believe that each of us sees and experiences the world in a unique way and at the same time deny that other animals see and experience it just as uniquely.

Cognitive dissonance is really just a name for the discomfort we feel when we both know something and avoid knowing it. In the case of where meat comes from, the result is a willful blind spot big enough to hide a mechanism of death so grisly, so gruesome, and so huge that it has already changed the face of the planet almost beyond recognition. And if we can miss several billion deaths without batting an eyelash, what else has been hidden in plain sight?

LET'S START BY looking at what's right at our feet. What we stand upon is the solid base of the food chain. It even shares the same name as our planet: we are sustained by earth. It is a scientific wonder, when you think about it: the colourful cornucopia of foods at your supermarket, all that variety—watermelons, strawberries, kale, hot peppers, spinach, lychees, Brussels sprouts, peaches, pumpkins, sweet potatoes—emerges from the same alchemy of water, sunshine, DNA, and, of course, dirt.

Soil health isn't easy to detect with the naked eye, but Canadian farmers have found an unusual technique to make it more visible: by burying men's cotton underwear in the earth for about a month, they can get an indicator of how healthy the soil is. That's because cotton underwear is 99 percent cellulose, essentially long chains of glucose molecules that provide a spectacular feast for the microbial residents of the soil. By placing several pairs of men's briefs in plots of land that have been farmed differently, you can get a good comparative gauge of how microbially rich the soil is.

Claire Coombs, a research technician at the Ontario Ministry of Agriculture, Food and Rural Affairs, put this to the test and buried several pairs of underwear in a conventionally tilled field grown continuously with soybean and in a no-till field with excellent crop rotation to see if there was a difference. After two months, the underwear in the tilled field was basically intact and wearable. The underwear in the no-till, high rotation field, however, was skimpy, to say the least. To hold it up, there was only the elastic band, the rest was "almost non-existent." The microbes had devoured the underwear, indicating that the soil was buzzing with life, a boon both for the earth and for the crops above.

It's been said that "civilizations rise and fall on the quality of their soil," and with our human population expected to rise to ten billion by mid-century, ignoring the steep degradation of soil would be a huge oversight. Citing a World Resources Institute report, British author Nafeez Ahmed writes,

> Over the past 40 years, about 2 billion hectares of soil—equivalent to 15% of the Earth's land area (an area larger than the United States and Mexico combined)—have been degraded through human activities, and about 30% of the world's cropland have become unproductive.[4] But it takes on average a whole century just to generate a single millimetre of topsoil lost to erosion.
>
> Soil is therefore, effectively, a non-renewable but rapidly depleting resource.
>
> We are running out of time. Within just 12 years, the report says, conservative estimates suggest that high water stress will afflict all the main food basket regions in North and South America, west and east Africa, central Europe and Russia, as well as the Middle East, south and south-east Asia.

---

4 According to the Food and Agriculture Organization of the United Nations, 25 percent of the world's land is degraded, but in areas like sub-Saharan Africa, Southern America, Southeast Asia and Northern Europe soil quality restraints affect more than half the land usage.

There is no single cause for soil degradation. It stems from drought and spreads with water and wind erosion due to lack of vegetation. But it also comes from industrial agriculture: from the dramatic rise of monocrop cultures, to the excessive use, and even lack of use of fertilizers. Ultimately, scientists have issued dire warnings that how we care or don't care for the dirt at our feet will soon affect two-fifths of the human population. The situation is so critical, a senior official at the UN has stated, that, at the current rate of soil degradation, we could have only sixty harvests left.

Soil is like a womb for seeds. It nurtures and nourishes plant life so that it can flourish and grow. But despite the fact that there are millions of different types of seeds (the renowned Svalbard Global Seed Vault, for example, currently contains 890,000 samples and has room for 4.5 million crop varieties), globally only twelve plant species and five animal species make up three-quarters of all our food.

And relying on so few species of plant or animal for food, and in particular relying on just one variety of a species, means that it takes only one disease or one major weather event to potentially wipe out a food source. It has happened before. In the 1800s, the Irish Potato Famine was caused by a water mould called *Phytophthora infestans*. The peasants in Ireland had been squeezed off the land as the landowners wanted to graze cows for beef, so instead they became dependent on monocropping one variety of potato called the "lumper." When the "blight" hit in 1845, the primary source of food for three million people turned into a rotting black slime.

The result was not just hunger, but one of history's most haunting tragedies. Over a decade, Ireland lost about 1.5 million people to death and emigration, about a quarter of its population. It took over a century to recover.

Bananas are just as vulnerable as potatoes. The Gros Michel

was the gold standard of bananas until the 1950s, when a fungus called Panama disease destroyed commercial crops. To create a scalable and identical product with no diversity, the seedless banana plants had been propagated by replanting cuttings, which meant that they were genetically identical clones. In fact, collectively, bananas are the world's largest single organism. And while most people have never tasted a Gros Michel, it's been said that we don't know what we're missing, because the Gros Michel was apparently much better tasting than the Cavendish, the type of banana we now buy in the stores. The Cavendish represents 99 percent of all banana exports, and as a seedless clone it is also under threat: a new, deadlier strain of Panama disease has spread from Asia to Africa and India, and is en route to Central America. When it arrives, many varieties of banana, including the Cavendish, could be wiped out.

But biodiversity is not just disappearing because of the plants we grow for our own food. According to the World Wildlife Fund, 60 percent of global biodiversity loss is due to land being used to feed our food.[5] That is, the land is used to produce feed for animals raised for their meat. And the way we "grow" meat is similar to how we grow monocrops: we control the seeds.

Like banana plants, many of our domesticated animals don't have sex. We control their gene pool too. If you trace your brunch back to its very beginnings, you'll find that natural sex

---

5 In the deforested Amazon, 80 percent of the soy grown becomes animal feed. The cleared land is also used for pasture for beef cattle. Exotic animals like jaguars, sloths, and anteaters are disappearing in the region, as over 700,000 hectares of their forest home was cleared between 2011 and 2015. For every hamburger and chicken wing there is corresponding loss of wild animals that inhabited ecologically rich areas like the Congo, the Amazon rainforest and the Himalayas. The link, once made, becomes obvious. To create the unnatural, we're destroying the natural.

has been snipped out of the picture. Your average dairy cow will be impregnated every year but will likely never see a bull in her lifetime. That's because today, for 95 percent of dairy cows and 90 percent of pigs, life starts not with a twinkle in an animal's eye but in a petri dish. The vast majority of these animals are conceived by artificial insemination.

For semen collection, the bull is either teased by a steer—a castrated male, which the bull will attempt to mount—or it is placed behind a dummy cow, which, in its simplest form, looks much like a vaulting horse from a high school gym. "Milking" the bull generally takes one of three forms. The first, and most common, method involves an artificial vagina. As the bull rears up to mount, a "collector" rushes in to fit the vagina over its penis, which is quite the job considering bull erections can be two feet long. To simulate the real thing, the rubber interior of the vagina is lined with a lubricant and its wall is filled with warm water. Using "thermal and manual stimulation," the collector harvests the bull semen. In artificial insemination centres, the bulls go through this collection process two to three days a week, and ejaculate is collected on those days two to three times.

For bulls that can't mount or are more difficult to restrain, the electro-ejaculation method is the preferred choice. The animal is guided into a metal chute so that only its rear end is accessible. The collector massages the bull rectally with a gloved hand to relax it. Next a large metal anal probe with electrodes is inserted into the rectum and delivers increasing pulses of electricity to the pelvic nerves. Vets use electro-ejaculation on wild and endangered animals to obtain semen and increase their numbers as well, though wild animals are anesthetized for the procedure. Bulls don't have that benefit.

The final method requires a more direct form of human involvement. The collector inserts a gloved arm elbow-deep into the bull's rectum, and massages the ampulla and accessory glands through the rectal wall, essentially masturbating the bull from the inside, until he ejaculates.

Usually, we think of cows as the ones we milk for profit, but ounce per ounce, bull sperm is far more valuable. A single straw of semen can go for as high as $2,000, with one ejaculation yielding up to five hundred straws.[6] At the elite level, one "star" bull can bring in over $7 million annually and sire over five hundred thousand offspring. As the lynchpin of the $600-billion global dairy industry, bull semen is considered "white gold." The bulls that sire the best milking daughters make fortunes for their owners and go on to become celebrity studs in the field. In the dairy industry, Comestar Leader, Sunny Boy, and Toystory are all household names. These bulls belong to the "millionaire's club,"[7] not only for the cash they bring in but specifically for their level of production: the ability to produce over a million doses of sperm.

---

6 There is variability in price. At the low end bull sperm can go for $5-$15 per straw, at the high end with Wagyu bulls the price can be $2000 per straw.

7 Obituaries are even written for famed bulls.

November 1, 1990 - October 24, 2005

Internationally respected, 71HO1181 Comestar Leader EX EXTRA was considered a true dairyman's bull. He left the mark of dairyness from head to tail on his daughters, siring high milk producers, show winners and highly classified progeny, leaving breeders satisfied around the world.

"This bull produced tremendous dairy strength with great udder texture and bone quality," said Lowell Lindsay, Semex Sire Analyst. "A Blackstar x Sheik x Mark Anthony, he was similar to his three full sisters for width of chest, great feet & legs and like his sisters, he passed it along to his daughters."

Leader's impact on the industry will be felt for generations to come as he had over 20,000 daughters classified in Canada alone. Certainly an impressive statistic on its own, but couple that will the fact that 67% of these daughters are classified GP or better, (270 EX and 3,411 VG) and it is easy to see why Leader will remain a favorite wherever dairy cows are appreciated worldwide.

So why is sperm such a big business? As geneticist Christine Baes explains, "A cow produces no milk unless she's recently given birth, which is why it's important to impregnate a cow as often as she can physically bear it—say, about once a year. A cow will milk for 305 days, rest for 60, and then be up to bat again." You'll note that unlike bulls, cows, who actually produce the milk the industry depends on, receive no public recognition at all. They are invisible. After a lifetime of producing milk—about 120 glasses a day per cow—they are considered spent and taken to slaughter to be ground up as dog food or hamburger.

This "jizz biz" is not just domestic; the United States and Canada export hundreds of millions of dollars' worth of semen straws around the world every year. Despite sanctions, the United States even exported $2-million worth of bull sperm to Iran, as it qualified as "humanitarian aid." Sending live animals in a 747 cargo plane is of course expensive, so instead semen is sent over in flat packs.

These days, with Nasdaq and the Dow Jones, we tend to forget that the stock market began as the trade in animal livestock. Today, however, even simple animal auctions have become high-tech and abstract: instead of putting live animals on the block, companies like Genomix host semen and embryo auctions. The goal is to buy the perfect line of genetics. Buyers look for traits like pounds of beef per cow, animal growth, and calving ease and check off the genes they want to boost their herd. In fact, bull semen has become such a hot commodity that in recent years there have even been barnyard break-ins, with thieves stealing semen vials worth tens of thousands of dollars on the black market.

This white gold is precious, not only for private business but also the state. Cryopreserved, or frozen, semen is stored in liquid nitrogen at –196°C. This allows it to be kept for at least fifty years, or even, some say, indefinitely. In the event of a large-scale disaster or disease, the U.S. Department of Agriculture has a secret liquid nitrogen storage facility, like the seed vaults for plants, to serve as a Plan B. Housed in Fort Collins, Colorado, the National Animal Germplasm Program (NAGP) is like a genetic Noah's Ark. The idea is that if something wiped out full populations of domestic animals on Earth, the NAGP has the ability to "repopulate entire breeds." More than seven hundred thousand straws of semen from eighteen different animal species are kept there at the ready. Semen of vintage, or "heirloom," breeds is also stored in the ten-thousand-square-foot facility alongside that of common breeds of pig, turkey, chicken, and cattle.

Maintaining these diverse genetic lines is important for another reason: to prevent inbreeding, a hazard of artificial insemination. With sixteen thousand daughters, five hundred thousand granddaughters, and over two million great-granddaughters, a bull named Pawnee Farm Arlinda Chief was once the Genghis Khan of the dairy world. Today, his genes can be found in 14 percent of all Holstein cattle. But farmers ended up breeding bulls and heifers together that were both descendants of the Chief, and Chief happened to have a faulty gene. If both animals carried a copy of it, it resulted in spontaneous abortions. For the industry, the financial loss was over $420 million.

Quality control, then, is a vital part of ensuring healthy sperm. At semen collection centres, the sperm are analyzed under a microscope to ensure that they are concentrated in good numbers, swim well, and have no physical abnormalities. Companies like Semex also use "computer assisted semen assessment systems," which use software and high-resolution video imaging to evaluate sperm according to company parameters. But in some places quality control is still done the old-fashioned way: along with sight, the semen is evaluated by smell.

While their job profile is likely not on LinkedIn, employees of companies like Finnpig have to sniff sperm and mark it for rejection if something smells off.[8] It's an important part of pig husbandry. Sabrina Estabrook-Russett, a veterinary student at the University of Edinburgh, detailed her experience as a farmhand, in *Modern Farmer.* Working for a Slovenian farmer she learned the customary method for testing the quality of semen. He told her: "We test by ALL the senses: see, touch, smell, taste. . . . When boar young, semen sweet. When boar old, semen bitter." Thankfully, she was spared the job of taste testing.

As a product, sperm is also offered to buyers in various selections. Sexed sperm, for example, is now a common commodity. For end-customers who want females to produce milk, it makes sense to buy XX sperm. Young males are unwanted, unless you happen to be in the veal industry. A cytometer is used to separate the male and female chromosomes by weight, and a magnetic current divides the XX from the XY sperm, so that you can buy sperm that will birth the required sex 90 percent of the time.

Some semen is even sold as "robot ready," meaning the daughters sired have teats that are more able to endure robotic milking machines. Dairy cows are known to get production-related diseases like mastitis, which is an udder infection. To address this, companies are changing the cow instead of the technology. As the Semex press release states: "Semex's Robot Ready™ sires will help dairymen make breeding profitable cows in automated, robotic dairies easy. . . . We've identified the need to produce cows that are suited for these systems as an essential requirement for our clients that are either already

---

8 "Our way of working with this problem has been careful to favor the middle ground. Seed taken from the herd is looked at what it looks like and how it smells. Any deviation in color, composition or smell is interpreted as abnormal and the item is rejected."

utilizing, or are considering implementing, automatic robotic technologies on their dairies."

By hijacking the biology of domestic animals, we have not only taken away their ability to have sex, we now have a hand on the dial that raises their numbers as well. From a business standpoint, there's a financial incentive to increase the product population. Pigs, for instance, are considered "mortgage lifters," and by increasing their offspring, you increase your profits. The rate of increase grows yearly by a factor of ten. Year one, 1 sow produces 20 pigs; year two, 10 sows produce 200 pigs; year three, 100 sows produce 2,000 pigs; at that rate of increase a farmer can be up to 2 million pigs by year six. But there are consequences: in the 1990s the average pig gave birth to 20 piglets a year. Today, with selective breeding, the rate has increased to 25 to 30 piglets, with some sows even giving birth to 40 piglets a year.[9] This rapid rise in animal production from artificial insemination and embryo sales has changed the process of reproduction, and as a result has dramatically increased the biomass of domesticated animals on the planet.

Today, there are over one billion domesticated pigs on Earth, one and a half billion domesticated cows,[10] and, according to annual slaughter numbers by the UN's Food and Agriculture Organization, almost sixty-six billion chickens. What this means, as George Musser, an editor at *Scientific American*, put it, is that "almost every vertebrate animal on earth is either a human or a farm animal." Including horses, sheep, goats, and our pets,

---

9 The high birth rate in pigs is leading to an increased mortality rate for the sows, as the litter rate has been "linked to a troubling rise in prolapse—the collapse of the animal's rectum, vagina, or uterus".

10 According to data from the UNFCCC, "If cattle were a country, they would rank third in greenhouse gas emissions."

65 percent of Earth's biomass is domestic animals, 32 percent is human beings, and only 3 percent is animals living in the wild.

The human population, currently at 7.5 billion, is rising at a rate of 1.2 percent per year. Livestock numbers are double that, at 2.4 percent per year. With our population expected to reach 10 billion by the middle of the twenty-first century, we will need to support not only 120 million metric tons of additional human beings, but 400 million additional metric tons of farmed animals. By 2050, the physical space required to raise food solely for livestock is expected to rise from three-quarters of all current agricultural land to half of all land in existence that is farmable, period.

A JELLYFISH AND A CUCUMBER are both 95 percent water. Humans are about 60 percent. What all terrestrial plants and animals have in common is that water is a part of our bodies; it makes up our food, and it quenches our thirst. We know we cannot live without it. But how much do we know about where it comes from?

Water, of course, is high above us in the skies. That's because clouds, in essence, are floating rivers. And while they look lighter than air, they do have a weight. The water content of the average cumulus cloud, for instance, is about 1.1 million pounds (495,000 litres) of water, or the weight of a hundred elephants. Clouds may be everywhere, but they are fickle in where and when they release their rain. The water we tend to rely on—for industrial, manufacturing, agricultural, and residential purposes—comes from two primary sources: aquifers deep underground, and snow and ice melt from glaciers. Both sources are disappearing.

To begin, let's start at the top, at the mountain peaks. All that glacial water is the origin of our rivers and streams. According to the U.S. Geological Survey, "runoff from snowpack alone provides 60 to 80 percent of the annual water supply for 70 million people in the American West."[11] That is why the photos of vanishing glaciers should be so alarming. In a warming world, without that annual snowpack cover turning into ice and then water, the runoff is not being replenished.

The rate at which this is happening is staggering. In British Columbia, glaciers are losing twenty-two billion cubic metres of water annually. That's like twenty-two thousand Empire State Buildings' worth of water vanishing from the mountain peaks every single year. In High Asia, mountaineer David Breashears has been documenting the loss of "frozen reservoirs." As the co-founder of the Glacier Research Imaging Project (GRIP), he and fellow mountaineers use archival photos to retrace the steps of mountain climbers taken over the past century to examine before-and-after photographs of the retreating glaciers. He writes,

> We should be uneasy. The loss of these frozen reservoirs of water will have a huge impact, as the glaciers provide seasonal flows to nearly every major river system in Asia. From the Indus, Ganges, and Brahmaputra in South Asia, to the Yellow and Yangtze Rivers in China, hundreds of millions of people are partially dependent on this vast arc of high-altitude glaciers for water. As the glaciers recede and release stored water, flows will temporarily increase. But once these ice reservoirs are spent, the water supply for a sprawling, overpopulated continent will be threatened, and the impacts on water resources and food security could be dire.

---

11 "In some regions, the vast majority of water feeding high-mountain rivers comes from rain and snowmelt, not melting glaciers." Some regions however, are dependent upon glacial run-off.

In 2016, the Center for Investigative Reporting began reviewing classified cables sent between US diplomats that had been released by WikiLeaks. The cables "showed mounting concern by global political and business leaders that water shortages could spark unrest across the world." Privately, the world's largest food company, Nestlé, has calculated that if everyone on Earth ate like the average American, the planet would have run out of fresh water fifteen years ago. Now, as populous countries like India and China play economic catch-up, their demand for meat has soared, and this, in combination with the trend of shrinking glaciers and dwindling aquifers, looks to be leading to a "potentially catastrophic" situation for Earth's water resources.

The problem is that the threats are invisible. Depleting groundwater used for crops has been called an "out-of-sight" crisis. Ancient rainwaters that have welled up in underground aquifers for thousands of years are being pumped out at unprecedented rates. And once this water is gone, it will take thousands of years to replace. We tend to think of water as coming from rain or melted snow and ice, but in fact we largely rely on this "fossil" water for modern agriculture. As Tom Philpott of *Mother Jones* magazine has observed, "To live off surface water is to live off your paycheck. . . . To rely on groundwater, though, is to live off of savings."

Right now, globally, a third of groundwater is in distress. And this groundwater blind spot looms under some of the world's most populous cities. There is however a way to see what's happening underground—rather surprisingly, it's by using satellites. NASA's GRACE-FO mission uses two satellites in the same orbit that follow each other. By constantly measuring the distance between themselves, they can detect changes in the gravity field they are sweeping over. And because oscillations of groundwater change the gravity field, the data generated can be used by scientists to "see" the volume of water that lies beneath. What they've discovered is that in some regions like California's

Central Valley, a volume of water depletion that used to take decades is now occurring over three years.

Our species uses about 4,600 cubic kilometres of water every year, double the volume of all the planet's rivers. According to the UN, by 2050 five billion of us will suffer water shortages; by 2025, just a few years from now, 1.8 billion of us will experience "absolute water scarcity." As a result, the cost of water is going up. A 2017 study by scientists at Michigan State University found that water is about to get a lot more expensive. They estimated that a full third of the US population would be unable to afford their water bills in five years.

Naturally, there is always surface water. Passing weather fronts bring the rain. And rain is free. But the water cycle is changing. As global temperatures increase year over year, warm air is causing more evaporation, leading to droughts in some areas and catastrophic floods in others. The cities that plan to use buildings and infrastructure to collect rainwater may well be the ones with drinking water in the years ahead.

WATER MAY BE SCARCE for humans, but it has always been plentiful for fish. Fish populations however, face a different menace. In recent years, they have come across an unrivalled predator, one whose hunting ability has become so refined it outpaces marine species' ability to repopulate. I'm talking of course about us.

In 1920, we developed a new and unusual way to fish: from the sky. As an article in *Aerial Age Weekly* noted at the time, the practice was seen in Virginia, where "each morning at 5 o'clock a flying boat carrying a pilot, radio operator and fish spotter leaves the station to aid fishing craft." By 1940, this "new use for airplanes" allowed fish spotters to track whole schools of fish from an altitude of about six hundred to eight hundred feet.

By the 1970s, spotter planes were commonly used by commercial fleets. Large catches became dependent upon this new way of seeing the fish. Unsurprisingly, a study submitted to the National Marine Fisheries Service in the United States found that 92 percent of fishing vessels that used aircraft had greater catch success.

For the fish, this technique has been devastating. With our eyes in the skies, they simply have no escape. In the Mediterranean at least, aerial spotting for bluefin tuna has been banned. After fishing fleets from Spain, France, Italy, Japan, and Libya came in using sonar and aircraft, the fish, encircled by the high-tech fleets and reeled in, really didn't have a chance and their populations plummeted. But that's the key thing. With fish being harder to catch because they are less plentiful, we need all the resources that we can to catch what's left. But our vision of what a normal population is changes. This is what scientists call the "shifting baseline." In the case of Pacific bluefin tuna, Kazuto Doi, a Japanese fisherman notes that, "Twenty years ago, we used to see the tuna swimming under our boats in schools that went on for two miles . . . we never see that now." That's because the Pacific bluefin population is now a mere 4 percent of historic levels. According to the UN, "nearly 90% of the world's marine fish stocks are now fully exploited, overexploited or depleted." An essential part of the diet for billions of people around the world is disappearing and most of us don't even notice. In fact, our demand for fish is increasing.

To cope with the demand, we not only catch wild fish species but also farm fish as well. That too has deleterious effects. Crammed into cages, fish can become diseased, covered in lice, or deformed. Fish health inspectors examining farmed salmon in Scotland, regularly find evidence of "bloody lesions, eye damage, deformed organs, plagues of flesh-eating sea lice" and more. According to the lobby group Scottish Salmon Watch, "The mortality rate on Scottish salmon farms is 26.7%." In other words, the process of farming itself kills fifteen to twenty million of the reared fish.

Then there is the fish that we catch that we don't intend to eat at all. Marine bycatch, or "fish waste"—the fish hauled in that are too small or are not the target species or sex—is sold on to factory farms as animal feed. Far from being a mere by-product of the fishing industry, fishmeal—made from small, wild-caught, bony fish—accounts for a whopping 60 percent of the global catch, making it easily the biggest part of the fishing industry, but one that the average consumer knows the least about. Every year, 5.4 million metric tons of these "trash" fish are caught, ground down into a powder form, and sold primarily as a supplement in factory farm feed, as a cheap source of protein.

In an increasingly overfished ocean, however, fishing fleets often move into illegal grounds. In Thailand's marine parks, tropical fish are commonly harvested and ground into fish flour to feed tiger prawns. These are the same prawns we find on our dinner plates in Europe and North America. The film *Grinding Nemo* documents trawlers netting up to fifty different species in the parks. Fishmeal is being ground up from colourful reef fish, sea horses, and endangered baby sharks, among other things. As the small fish vanish, this reverberates up the food chain; the juvenile fish don't grow to be adults, and fish that are typically hunted by larger marine predators are removed entirely, leaving those predators with precious little to eat.

Peru is the largest fishmeal-producing country in the world. And a third of the country's catch goes to raising Norwegian farmed salmon. To produce one kilogram of farmed salmon requires two to five kilograms of feed made from small fish. In Peru, fish that could be feeding locals is instead being exported.[12]

---

12 One of these is anchoret, a species of anchovy, which have fed Peruvians for thousands of years.

Similarly, in West Africa, fishmeal plants have begun popping up along the coasts of Senegal and Mauritania. In Senegal, monster trawlers have slashed the fish biomass from one million metric tons to four hundred thousand. Beyond the horizon of our sight, our local supermarket chickens are being fattened up with fish from Africa. For West Africans who once lived along a coast with one of the richest fisheries, it's getting harder to buy and eat fish from their own waters, because their fish are being shipped overseas to feed our chickens.

IN THE 1960S, 80 percent of chickens were sold to the public as the whole, recognizable animal. Which is also why, at the time, chicken was not a popular meat. Dressing, cooking, and carving a whole bird was time consuming, so chicken was primarily reserved for Sunday dinners or special occasions. All of this changed however when Robert Baker, a professor of food science and marketing at Cornell University, developed the first machine for stripping meat from a chicken carcass.

Known as the Thomas Edison of the chicken industry, Baker also helped invent the deboning machine, and his pioneering work on binding agents allowed the flesh and gristle to stick together, creating a new commercial market for processed meat. This "fun" meat could be formed into kid-friendly shapes, like stars or hearts or even dinosaurs.

The transformation of chicken from whole bird to more commercially saleable parts—like drumsticks, wings, thighs, and breast meat—and processed shapes hugely expanded sales. The $4-billion-a-year industry of the 1960s was revolutionized. Demand for chicken meat soared, and today over sixty billion chickens are slaughtered every single year, the vast majority of which, 75 percent, come from factory farms.

Along with a sustained marketing blitz, chickens were no

longer just birds but brands and commodities. And with increasing demand, meat processing plants had to ramp up supply. To do this, the slaughter line became increasingly automated, and animal *disassembly* sped up. Today, this is the last stop for a commercially raised hen: the bird's life reaches its end not at the hands of a human but with the blade of a machine.

With that, we have opened up a truly gruesome blind spot. Let us bravely proceed.

On the kill floor of a slaughterhouse, the "live hangers" are the workers who hang chickens upside down in stainless steel shackles. The line moves quickly. To keep pace, a worker will need to hang up to twenty birds a minute on average. Then the machine takes over. Like a cross between a horror film and a theme park ride, the birds are cranked along a rail and over to a water bath, where they are stunned as their heads are electrocuted underwater.

The killing machine is next. Each bird lands in a position with its neck fixed between guide bars. Then a motorized circular blade moves in to slit the bird's throat. As the birds bleed out, a camera-based counting system tracks the number of birds that go by. Just after the kill, the birds enter another high voltage chamber. This one is called a "stimulator." Here, the hens are run along an electrode plate for approximately forty seconds. This induces breast contractions and a flapping motion in the birds to exhaust any chemical energy left in the muscle. Tenderizing the breast muscles makes the deboning process faster and easier.

Next, the hanging bodies enter scalding tunnels, the longest part of the line, where heat from hot water is transferred to the birds' feather follicles so that they can be plucked by the scaler at the next stop. Scalding is either "soft" or "hard." Soft scalding, at 55°C, produces yellow-skinned birds, and hard scalding upwards of 57°C produces birds with white skin.

The conveyor then lifts the birds back up through the final stages of the operation: the head puller, where the birds are decapitated, and the hock and feet cutter, where the birds' legs and feet are chopped off and they are released from the shackles.

At this stage, except for legs and head, the birds are still whole. They still need to be eviscerated, chilled, and inspected before being rehung on the line. Then the carcasses are cut in two crosswise on a halving wheel. The "saddles," or back halves, which include the thighs and drumsticks, are packaged in what's sometimes called the "leg room," while the front halves, with the breasts and wings, are split up by a harvesting machine. Modern robotic deboning machines can process one bird in 2.5 seconds.

In Europe, Asia, and Canada, processing plants run line speeds from 175 to 200 bpm (birds per minute). In a single plant, that amounts to twelve thousand birds per hour, or ninety-six thousand birds in an eight-hour day. In the United States alone, nearly nine billion broiler hens are raised for slaughter every year.

Speeding up slaughter to maximize efficiency is not just happening in poultry processing, however. Meat is big business. The trend is industry-wide, with chickens, pigs, and cows—whole animal farms by the day—all barrelling down the line. And these animals are not the only casualties. At US meat plants, the increase in line speed is causing serious injuries to workers. According to records from the Occupational Safety and Health Administration, worker amputations happen on average twice a week.

So how did we go from farm to factory? It was a Chicago meatpacker, Gustavus Franklin Swift, who first invented the conveyor belt system used in meat processing plants. A shrewd businessman who was always looking to maximize his gains, Swift saw the inefficiency of transporting whole animals from Chicago to other major US cities by rail. At the time, 60 percent of an animal's mass was considered inedible. Heads, hooves, bones, and innards were all dead weight that needlessly contributed to shipping costs. Swift's idea was to take larger animals like hogs and cattle and dismember them in advance. By cutting hogs up at the stockyards, for example, he could transport them across the country as hams, ribs, bacon, and sausages using refrigerated rail cars.

Swift's contemporary Philip Danforth Armour, the founder of Armour & Company, found a different way to transform animal carcasses into something more profitable. He made his fortune by packaging the first shelf-stable meats in the form of canned chilis, hashes, and stews. But Armour also maximized the economic value of animals by finding new uses for the slaughterhouse waste products: pigtails became paintbrushes; bristles became hairbrushes; guts became tennis racket strings; fats were made into soaps; bones became fertilizer; and hooves were boiled up into glue.

Chicago's Union Stock Yards became a centre of "remorseless innovation." Armour himself famously proclaimed that he sold "everything but the squeal." Today, the meat industry despises the term "factory farm," but the cold fact is that animals are regarded in every way as commodities: they are raised as "products" and sold in "units."[13] As Ted Genoways writes in *The Chain*, the whole model is "carried out with the exactitude of a factory, and built around serving the needs of other factories—the packinghouses, the packers, shipping warehouses—farther along the supply chain. Each step can be replicated and repeated countless times in identical or near-identical facilities almost anywhere the residents of nearby communities will allow it."

While the slaughter of chickens has been fully automated for some time, for larger animals like pigs and cows, which are of

---

13 In Iowa, for example, livestock are sold by "animal units" not by headcount and are measured by weight according the equivalent of standard-sized cattle. A hog, then, is considered to be 0.4 of an animal unit.

higher value, death is still largely a manual affair. In most meat plants, restrained or stunned animals still have their throats slashed manually. As the blood drains out, it's collected in a trough and rendered to become blood meal. Along with ground-up bone, it is repackaged and sold as bone meal, a rather squeamish thought for vegetarians, who may forget that growers of everything from corn crops to vegetable gardens use animal blood as "organic fertilizer."

Sweet treats like gummy bears, candies, marshmallows, and Jell-O are also slaughterhouse by-products in disguise. Their key ingredient is gelatin, made from skin, bones, horns, and connective tissues collected after slaughter and placed in lime slurry pits for about three months to loosen the collagen in them. The lime is then washed off and the collagen is boiled down and turned into gel sheets or powder used in almost every moulded dessert. But gelatin's binding power is not only used in food; it has become a valuable commodity in everything from the soft pill capsules used in pharmaceuticals to paper production. All photographic film is made using gelatin. That's what the "film" that coats the plastic base *is*. The gelatin is the medium that suspends silver halide crystals that react to light. Which means every movie that was shot on film—from *Star Wars* to *Lord of the Rings*—was projected through slaughterhouse remains.

To insiders, it's known as "the invisible industry." Rendering, the process of transforming inedible animal parts into other products, is now a multi-billion-dollar business. In North America alone, over sixty billion pounds of "animal waste" are turned into commercial goods every year. In the 1940s, rendered animal parts could be found in around seventy-five commercial products. Today, the list is incalculably longer: antifreeze, cement, bullets, waterproofing agents, fabric softeners, detergents, chewing gum, fireworks, sheetrock, plywood, crayons, paint, insulation, and linoleum are just a few of the everyday items that contain unrecognizable animal remains.

Animal parts also become feed for the pet food industry. Globally, pet food sales are soaring. It's a $66-billion-a-year industry. But no one stops to ask which animals the "meat" comes from? Along with slaughterhouse slurry, or meat meal—which is made up of the parts we don't like to eat, like eyes, feet, and brains—other forms of "mystery meat" have been known to make their way into feed for our pets. Researchers at Chapman University in California examined the DNA in fifty-two different commercial pet foods and found that sixteen of them contained meat from a species that was not on the label. According to an article in *Modern Farmer* magazine, the pet food industry has a dirty secret. Producers have been found purchasing meat meal "from rendering plants known to accept euthanized shelter animals. Other reports have disclosed the inclusion of roadkill, restaurant fryer grease, spoiled supermarket meat, and the remains of diseased zoo animals."

Over seventy billion animals a year meet the grim reaper in an industrialized setting, but what's equally shocking is how invisible all this killing is. After the Second World War and the Holocaust, we may have thought that the grotesque horrors of gas chambers had been ended, but for animals the method was reintroduced in the 1980s and '90s, and gas chambers are widely used to this day. Controlled atmospheric stunning (CAS) is considered a humane method for rendering pigs and poultry insensible before slaughter. But inside the gas chambers themselves there's incredible suffering. Animals gasp, shriek, convulse, and attempt to escape as CO_2 increases in the chamber. In electric stunning systems for poultry, some birds emerge from the water bath still conscious, and, because of increasingly fast line speeds, birds that are not properly attached by workers to shackles sometimes miss the automatic blade that is meant to kill them. As a result, in the United States anywhere from seven hundred thousand to a million birds a year are still conscious when they are scalded to death in the scalder.

Civilization and polite society have raised us to believe we are above the barbarism of beasts. But that is hardly so. Our clinical methods of killing have simply allowed us to turn a blind eye to the horrors of food production. As the British author George Monbiot writes, "What madness of our times will revolt our descendants? There are plenty to choose from. But one of them, I believe, will be the mass incarceration of animals, to enable us to eat their flesh or eggs or drink their milk. While we call ourselves animal lovers, and lavish kindness on our dogs and cats, we inflict brutal deprivations on billions of animals that are just as capable of suffering. The hypocrisy is so rank that future generations will marvel at how we could have failed to see it."

5
BLACK GOLD

The obscure we see eventually,
the completely apparent takes longer.
—EDGAR R. MURROW

IT WAS ALL GOING ACCORDING TO PLAN until the baby pandas showed up. The National Grid operators had their eyes trained on their television sets as usual, but this time the expected spike did not occur. They were looking for a phenomenon known as TV pickup. In the United Kingdom, it happens when massive audiences watch sporting events like the World Cup, or the season finale of popular shows like *EastEnders*. As the program ends, the Brits have a tendency to all get up at the same time to take advantage of the commercial break and make a cup of tea. Behind the scenes though, this seemingly innocuous task sets off a cascade of events. As millions of kettles are switched on at the same moment, the sudden spike in demand for electricity sends a huge, sweeping surge across the grid.

When a massive jolt of power is required that's beyond the base capacity of the National Grid, its operators must turn to an additional source of power that is ready to go at short notice. But at peak capacity, you can't just "switch on" an additional power plant, because they are slow. If you think powering up a computer sucks up precious time, you'll appreciate it's a problem when fossil fuel generators take about half an hour to gear up, and nuclear power plants take even longer. During a commercial break and the national rush for the kettle, there's no time for this lag; it all has to happen immediately. So, the engineers turn to pumped-storage reservoirs. That is, at times when electricity demand is typically low, they pump water up to a reservoir at the top of a hill. Then, when demand spikes, they let it rush back downhill so that it spins a hydroelectric turbine along the way.

On this particular occasion, the credits had begun rolling for *The Great British Bake Off*, a hugely popular show, and so the grid control centre was at the ready, prepared as usual for the commercial break. But they had not anticipated what the BBC had programmed next: a nature documentary featuring baby pandas. As a result, nobody made tea, because nobody got up. According to a representative from the National Grid, "There was no pickup at all." The mass migration to the kitchen for a cup of tea had stalled, because people sat awed by the bundles of black and white cuteness and sat glued to their TV screens.

We don't tend to think of our electricity as being made fresh to order, but it is. It's made the moment we request it. When you charge your phone, as Gretchen Bakke writes in her book *The Grid*, the power you use is "so fresh, that less than a minute ago, if you live in wind farm territory, that electricity was a fast-moving gust of air. And if you live in coal country, it was a blast of pulverized coal dust being blown into a 'firebox'—a huge, industrial, flash-combusting furnace. If you live in hydro country it was a waiting rush of water dammed up by a massive concrete wall. Picture it. The electricity you are using right now was, about a second ago, a drop of water."

Most of us never think about electricity, let alone the grid. It is the biggest and most powerful machine in almost every nation, but we ignore it even when it's right in front of us.

In the late nineteenth and early twentieth centuries, the grid's relationship to our daily lives was a little harder to ignore. That's because utility companies' black, messy swaths of wire tangled up city centres like out-of-control spiderwebs. Not only were there the newly installed telephone and telegraph lines, but electricity cables were also attached to tall poles, and the wires were not neatly subdivided. Aesthetically and functionally, it was a nightmare. In London, the wires of sixty-five different electric utilities were strung above the streets. As technology historian Thomas Hughes writes in *Networks of Power*, "Londoners who could afford electricity toasted their bread in the morning with one kind, lit their offices with another, visited associates in nearby office buildings using still another variety, and walked home along streets that were illuminated by yet another kind."

In America, hundreds of patents were being issued for different types of dynamos used to generate direct current, or DC, electricity. In the 1890s, Chicago alone had forty-five different electrical utility companies and had dedicated DC power lines at 100, 110, 220, 500, 600, 1,200, and 2,000 volts. Needless to say, it was a mess.[1]

Today, we give little thought to how the power at our fingertips is delivered to us. We take for granted that it will be there when we press a button, flick a switch, or plug a cord into an outlet. This "normal," as Bakke notes, is a "blind luxury." Occasionally, in the street we catch sight of a notice—the photo of a missing cat, a flyer for yoga classes—stapled to one of the wooden poles that line our neighbourhoods, but seldom do our eyes follow the poles upward to the network of wires that criss-cross the urban landscape.

What we *do* notice is when the power is suddenly gone.

After the magnitude-9 earthquake and tsunami that struck Japan on March 11, 2011, the Fukushima Daiichi nuclear plant stopped producing power for the grid. When the quake hit, eleven of Japan's fifty-four reactors shut down, leaving the nation with a ten-million-kilowatt shortfall in power. Tokyo was instantly put on an energy diet, as rolling blackouts carefully rationed the power being used in the city. In some prefectures, power was out for up to six hours a day. Having no choice but to accommodate, factories shut down, restaurants closed, as there was no power to refrigerate or cook the food, people sat in the

---

1 It was Thomas Edison, famed inventor and creator of the first DC electric utility, who began tucking Manhattan's "black spaghetti" of wires away from sight. After much convincing, New York's mayor reluctantly allowed Edison to dig up the city streets and lay down eighty thousand feet of underground wiring to bring "electric light" to people's homes. Edison's utility project however did not last, however, as he was outdone by another famed inventor: Nikola Tesla. Tesla's alternating current (AC) power delivery systems could move electricity over much longer distances. This meant that generators weren't eyesores, and could be built far away from populated cities, but it also meant that our source of energy would be hidden from us.

dark, and up to half the trains stopped running. ATMs stopped working, escalators and elevators only ran sporadically, phones couldn't be charged, and without traffic lights there were a lot more car accidents. Even Tokyo's iconic illuminated billboards went dark in the heart of the city. In short, business could not go on as usual, because daily life relies so deeply on electricity. As NBC News reported, "one of the world's most technologically advanced societies was transformed overnight into one of Third World hardship." Without power, Japan simply fell to its knees.

It's not only large-scale natural disasters that can take out the grid. While our minds might leap to the threat of foreign hackers wreaking havoc on our power supply, a more innocent creature is responsible for most critical infrastructure blackouts. As John C. Inglis, former deputy director of the National Security Agency (NSA), has stated, "I don't think paralysis [of the electrical grid] is more likely by cyberattack than by natural disaster. And frankly the number-one threat experienced to date by the US electrical grid is squirrels."

Squirrel outages are typically short and limited to a single neighbourhood. A much bigger threat is trees. In 2003, the largest blackout in North America's history was caused when three overgrown trees brought down power lines in different parts of the grid, causing other lines to pick up the additional burden. Like electron dominoes, the blackout spread over 240,000 square kilometres, and for two days fifty million people in Canada and the United States lost power. It led to over $6 billion in lost business revenue and a dip in America's GDP.

In response to this, the Transmission Vegetation Management Program was formed, basically a regional trimming service to ensure that trees and tree branches don't come down on high power transmission lines. But to keep unsightly transmission lines and towers out of the way, they are often built over rugged and hard-to-reach terrain, which makes tree-trimming a difficult task. Covering the expansive territory by foot or ground vehicle alone is too slow, which is why today an incredibly dangerous occupation exists: the helicopter buzz saw operator. Officially known as "aerial side-trimmers," the choppers are outfitted with a forty-foot-long buzz saw rigged with ten circular blades that dangles vertically below the cockpit. The pilot has to deftly fly a path alongside the lines and towers, trimming trees that are growing too close.

This high-tech solution may solve one problem, but the much bigger challenge is that the grid itself is growing old and rickety. Today, 70 percent of the grid's transformers and transmission lines are over twenty-five years old, and because of the inefficiencies built into the older infrastructure, power outages are not only increasing but the time to get back online also grows longer year by year.[2] According to one estimate, to upgrade and replace this integrated network in the United States would cost over $5 *trillion*.

The mash-up of energy that goes into the grid is also a concern, because the infrastructure was originally built to deliver a steady output of energy from traditional nuclear, oil, coal, or gas powered plants. It was designed as a centralized form of energy delivery. But today, with the goal of adding renewables, we are hooking up many new ad hoc, decentralized sources of energy and feeding them into the grid. Wind, solar, and geothermal are some of the renewables we're familiar with, but these days energy can be made from just about anything—even cheese.

SAVOIE IS A PICTURESQUE REGION in the French Alps. Known for its ski hills and cozy villages in the winter and lush alpine

2 As Bakke notes, the average outage in the United Staes is 120 minutes and growing year by year; in other nations it's ten minutes and shrinking.

landscapes in the summer, it is also renowned for its Beaufort cheese. Two of the main by-products of the cheese-making process are cream and whey. The cream is transformed into products like butter and ricotta cheese, while the whey is skimmed to make whey protein powder for shakes and energy drinks. But the liquid leftover from the whey is not thrown out as waste; instead it's used to generate the town's electricity. In the town of Albertville, 1,500 people have their homes powered by cheese.

The secret to this power is micro-organisms known as archaea, which are added to the liquid in an oxygen-free, or anaerobic, digester. Here, they feed on the sugars in the whey liquid for four days, releasing microscopic burps in the form of biogases: carbon dioxide and methane. The gas is then purified and burned much like natural gas to heat water to a near boiling temperature of 90°C, which produces steam. The steam drives a turbine, which has a shaft connected to a magnet that spins rapidly inside tightly wound coils of wire. The magnet causes the electrons to get stripped away from the atoms making up the wire, and it's this magnetic force that physically creates electricity.

If you don't have a sufficient supply of cheese, you can always burn coal or oil to produce the steam that drives the turbine. Or boil water using the radioactive decay of uranium-235. (Or capture the kinetic energy of water rushing downhill, which, as we saw, is how the British power their kettles to make tea all at once.) But for the most part, the basic underlying mechanism is the same: we make electricity by turning a turbine.

Our civilization is powered by the invisible forces of electricity and magnetism. We all want it, we all demand it, but most of us don't even know what exactly it is. When you turn on a light—say a 120-watt light bulb in your bedside lamp that's drawing one amp of current through it—the equivalent of six quintillion electrons are zipping through a single point in the wire every second. But the electrons that run through your light did not make their way there in a flash direct from the power plant. The electrons themselves move slowly, it's the energy that moves quickly.[3] That's because electrons, as subatomic particles, do not flow down the wires like water in a pipe. The process works more like a wave. When you hear someone singing in the distance, the sound pressure that hits your eardrum isn't from the air molecules that came out of the singer's mouth. Sound is a compression wave. The air molecules bump into adjacent air molecules, rippling over the distance, and what you hear is like the last domino in a molecular domino effect, a reverberation of sound that comes from the singer.

In the same way, electricity is a ripple effect. Driving through a city at dusk, you've likely seen the street lights come on all at once. That's because the electricity doesn't leave the switch and make its way down the street. As soon as you add one electron to the wire at one end, another will pop out the other end. That is, at the atomic scale, when the generator cranks the magnet over the copper coil, ripping electrons away from the copper atoms, the now "homeless" electrons have to go somewhere, and where they'll go is over to the next available atom, joining its orbit. But in doing so, this will knock out the neighbouring atom's electron, which then bumps over to the next atom, and so on. But not all atoms are welcoming. Some materials—like rubber, for instance—lock their atomic doors to outside electrons,

---

3 You might be surprised to know that the electrons are moving down the wire incredibly slowly, as in — slower than tortoises. We're talking a drift speed of about 1 meter per hour. As tiny subatomic particles, electrons aren't orderly. They move in a haphazard way. And in an alternating current (AC), which is what is on the grid today, the electrons are constantly taking a few steps forward and a few steps back.

making it harder for electrons to move through them. We call these materials insulators. Metals, on the other hand, tend to have an open-door policy. They are conductors. Here, electrons can move about freely, making the jumping process swift and easy as they go from proverbial door to door.

And swift and easy it is. The average Canadian home uses about nine hundred kilowatt hours of electricity a month, to run the dishwasher, the dryer, the lights, the water heater, the air conditioner, the refrigerator, computers, televisions, and electronics. (Incidentally, that is one of the highest levels of consumption in the world, comfortably ahead of the United States.) But that still doesn't tell us much. Astrophysicist Adam Frank worked out another calculation, breaking down how much "pedal power" it would take to create electricity for the average home. To give you an idea of how much energy (or how little) the human body can generate, it would take fifty people pedalling eight hours a day for a month to power the average house. And just one person? Pedalling eight hours a day, the average person could generate enough power to light a single lamp.

While we're very good at generating electricity, one thing we have not been good at, until very recently at least, is storing it. So the next time you charge your cell phone, you might want to thank the torpedo fish. It's the reason your phone, and every other electronic gizmo you own, has a battery. The unusual fish, able to shock its prey with two hundred volts, fascinated an Italian physicist named Alessandro Volta, who in the 1790s set out to create artificial electricity to mimic its abilities. The fish, Volta noticed, had an organ with a particular pattern of chambers on its back: these four to five hundred columns were each filled with a stack of four hundred jelly-filled disks known as electroplaques.

Ever the experimenter, Volta combined this observation with something he'd discovered by chance: the taste of metal coins. He knew that if he placed coins made of different metals on his tongue and put a silver spoon on top of the stack, he could feel the weak but distinctive tingling of electricity. He wondered whether, if he stacked more of these metals together, like the torpedo fish had done naturally, he could generate more of this strange power.

His invention became known as the "pile," because that's exactly what it was: a pile of two different metals, copper and zinc, stacked up high like pancakes with a brine-soaked cloth in between each disk. By connecting a wire to the top and bottom of the pile and placing the ends again on his tongue, he found that a constant current, more powerful this time, was flowing through. He had invented the world's first battery.

Today, batteries are everywhere; we use them to power our electronics, which are so called because they are machines that run on electrons. And while we think about charging them every day, (consider the pangs of dread at the thought of a drained phone battery), few of us give much thought to how this power works.

The lithium-ion batteries for our electronics, or zinc-carbon, nickel-cadmium, or lead-acid batteries, all in essence work the same way. A battery requires two different metals, one that likes to "give" electrons and another that likes to "receive" electrons, which give the subatomic particles a direction to move, from givers to receivers. And while batteries that fit in our pockets are commonplace, the big quest today is to create massive, building-sized batteries capable of storing far larger amounts of energy, to provide a boost, when it's needed, on the grid.

As in Britain, when everyone gets up off the couch during the commercial break for a spot of tea, there are predictable periods in city life when power use surges. On hot summer afternoons in Los Angeles, for instance, the whole city relies on air conditioning, but there still needs to be power available when people come home from work, turn on their televisions, and start cooking dinner and using other appliances. Not every city has a pumped-storage reservoir to create an artificial waterfall to generate power. In LA, the city prepares for high-demand times by turning on a "peaker," a gas-burning power plant, to accommodate need. The fossil fuel plant dates back to the 1950s and is old and inefficient. So Los Angeles is switching things up, and by 2020 it plans to have the world's largest storage battery: a building filled with eighteen thousand lithium battery packs that can turn on in minutes, rather than hours, and give LA a boost for an additional four hours. In Australia, Elon Musk's big Tesla battery is already operational. Its first big test came in December 2017, when one thousand kilometres away a coal unit tripped, causing a slump in power. The battery kicked in within milliseconds, pumping 7.3 megawatts into the grid, much faster than a nearby coal generator could.

Whereas batteries of the past were big and clunky, too hefty even to put in cars, today they are light and small. We carry lighting in our pockets and have been able to miniaturize our cell phones and other electronics, while also making them more powerful, thanks to a new type of battery powered by a metal called lithium.

Lithium is the lightest metal on Earth. It was discovered by Soranus of Ephesus, a Greek physician in the second century AD, who would treat manic patients in the alkaline waters in his town. Even today, the same lithium that we use to power electric cars is used as a treatment for depression and bipolar disorder. Scientists know that the element affects serotonin levels in the brain, but they still aren't clear exactly how.

One of the biggest deposits of lithium is hidden under the largest natural mirror on Earth. Bolivia's Salar de Uyuni, the world's biggest salt flat, covers ten thousand square kilometres and gleams a striking reflection back up to the sky when the salt is covered by a thin layer of water. The mirror is so big, it can be seen from space. As for the lithium, ancient volcanoes deposited the metal into prehistoric lakes that evaporated, leaving the white crust of salt, with a pool of blue-green brine containing the "grey gold" lying five metres beneath the surface. According to the U.S. Geological Survey, the Salar contains 5.4 million metric tons of lithium, while the Bolivian government claims the number is much higher, at 100 million metric tons, or 70 percent of the world's reserves. But because of Bolivia's history, with foreign governments sweeping in and exploiting the country for its silver and tin, Bolivians are protective of their resource. Their goal is to start their own lithium mines, run by the people for the people, instead of letting big corporations in.

That said, lithium mining in Bolivia has been slow to get going. The lithium in your phone most likely comes from Australia, Chile, Argentina, or China,[4] countries that have been working strategically to corner the market. The United States also produces lithium, from brine pools in Nevada. About one gram of the metal can be found per litre of brine. For perspective, a cell phone battery contains about five to seven grams of lithium carbonate—the powdered form of lithium—but a lithium car battery requires up to thirty kilograms. For high-end electric cars like the Tesla Model S sedan, the amount is sixty-three kilograms of lithium carbonate, or the equivalent of the amount for about ten thousand cell phones.

IN 1905, A TWENTY-SIX-YEAR-OLD named Albert Einstein published the law of the photoelectric effect, essentially showing how light can create electricity. Two decades earlier, it had been observed that certain elements, like selenium, could generate

---

4 Lithium is also mined from hard rock pegmatite. This is a more traditional type of mining, sourcing the metal from ore, and is common in Australia and parts of China.

an electric current when exposed to light, but nobody knew why or how this was possible.

The prevailing theory at the time was that light was a wave. And if that were the case, then increasing the intensity of light should produce more electricity. But that's not what happened. Even weak light could loosen electrons from their orbit. Einstein's genius and insight was in postulating that light wasn't only a wave; it was also a particle. And these particles, or photons, acted like an eight ball in pool: if you had enough of them—that is, if there was a high enough frequency rather than intensity—they could knock an electron out of an atom's orbit and into space. At any given moment, there are a lot of photons hitting Earth. To get a sense of how many, consider that on a clear day a square metre of the planet's surface receives about one thousand watts of solar energy. Now, consider that Earth's area is over five hundred *trillion* square metres. There is plenty of solar power to go around.

The solar panels we use today are the direct result of Einstein's Nobel Prize–winning discovery. Photovoltaic cells use particles of light to knock electrons free from atoms to create an electric current. But here the electrons don't bounce out into their surroundings, they are kept inside of the semiconductor material. The current can then be used for power. As solar cells increase in efficiency and drop in price, they, along with lithium batteries to store the energy, are our greatest hope for humanity transitioning to non-polluting energy.

Demand for solar energy is soaring and installations have jumped by 50 percent worldwide. While this is very promising, solar, for now at least, still only powers a minuscule amount of energy for the grid. According to *The Guardian* newspaper, even in Europe, where solar power is most prevalent, it provides only 4 percent of electricity. The problem is that the sun can't be counted upon to be shine when everyone gets up to plug in their kettles. In fact, the sun shines when we tend to need it least. Particularly in northern climates, and particularly in winter, peak demand occurs when it is dark. Until we have batteries that can store the energy of sunlight—and until governments upgrade the grid (which was designed as a one-way system from power plants to homes) to a decentralized system that functions just as well with home solar installations feeding back into the grid—most of that clean solar power will continue to reflect back into space.[5]

A different way to harness the sun's power is by tapping into the wind, something people have been doing since the first century AD. The Netherlands, for example, is famous for its windmill technology, and by 1850 there were over ten thousand windmills dotting the Dutch countryside. As the sun heats Earth's surface, it creates warm air, which rises and leaves an area of low pressure. As a part of nature's balancing act, this causes other air molecules to rush in from cooler high-pressure areas. This swirling interaction is the invisible force of the wind. Windmills harness this power as it sweeps by, and the modern wind turbine can convert it to electricity.

But the wind is a fickle beast. It shows up on its own accord and it does not blow at a steady rate. Sometimes, it doesn't blow at all. The grid, on the other hand, requires a steady input, and a tidy balance between input and output. Not too much, and not too little. This causes obvious problems. If wind turbines don't spin, there is no power. And too much of a good thing is an even bigger problem. When the wind decides to go full force, things can also go haywire. As Gretchen Bakke writes, "You can't just turn the wind down. When it blows hard . . . you can see it in the

---

5 That is, minus all the energy that has been absorbed in photosynthesis, or the hydrological cycle, or the many other indispensable things the sun provides.

power spikes—*bang, bang, bang*—of wind farm after wind farm shooting electricity into the system. It floods the grid; it crashes through the infrastructure much like a wave crashing against a sea wall on a stormy day. Even Los Angeles can't absorb all the electricity made on a seriously blustery day in the Pacific North West. . . . When there is too much power on the wires they overload, or circuits break to protect them, and in so doing they close, rather than open, available paths for excess power to take."

When this happens, a blackout can result.

YOU HAVE TO WADE PAST many tourists with selfie sticks to find a spot to sit and relax, because it's an incredible draw. Every year, hundreds of thousands of tourists make their way to the southwest coast of Iceland to bathe under the midnight sun, sip cocktails, and soak in the country's most famous attraction, the Blue Lagoon. The setting is spectacular. Nestled among the lava plains, the cyan-blue waters steam in the crisp air, and the water's mix of silica, algae, and minerals is even said to have healing properties. But what surprises many is that the Blue Lagoon is not a natural hot spring. It may make the experience seem less mystical, but it's a human-made attraction that's fed by outflow water from the Svartsengi geothermal power plant next door.

Two kilometres below the surface, thirteen boreholes bring superheated groundwater that's been deposited near magma up to the surface. The country's volcanic terrain is what's allowed Iceland to access this powerful heat from Earth's core. The steam powers turbines that generate electricity and provide hot water for twenty-one thousand nearby households. Today, five major geothermal plants along with hydropower have made Iceland one of the few countries in the world that use renewable sources for 100 percent of their electricity. And while the industry is booming—forty countries are in geothermal-rich territory—the cost of drilling deep towards Earth's molten core and the dependency on location have meant that even today less than 1 percent of the world's electricity comes from geothermal power. That said, as technology advances, a larger role is possible. The World Energy Council estimates that in the future the number could be as high as 8 percent.

THE AWESOME POWER OF WATERFALLS is another alternative to fossil fuels.[6] In fact, the grid's beginnings were formed by a power plant situated in the "honeymoon capital of the world," Niagara Falls. Nature's torrential beauty has been harvested here since 1896, when the power of the water was first used to spin giant turbines to generate electricity. Niagara was the first place to use Nikola Tesla's alternating current (AC) power. By inventing what's known as "polyphase alternating current," Tesla was able to use timed sequences of electrical current to create a rotating magnetic field that could spin a motor. This new type of power meant that current could be sent over a distance in the wires, to move an object through magnetism on the other end. The invention was brilliant, and to those who saw it for the first time, it must have seemed like magic. Soon the roaring waterfalls brought electricity to the city of Buffalo, thirty-two kilometres to the south, and within a few short years Niagara Falls was powering the bright lights of New York City.[7]

---

6 Where nature's waterfalls are not available artificial falls are created in the form of engineered gradients where a drop in land elevation is used to channel rushing water down tunnels via gravity to achieve the same thing.

7 Today, Niagara Falls churns out almost 2 million kilowatts of power on the Canadian side, and 2.4 million kilowatts of power on the US side.

Tesla's invention is still considered one of the greatest of all time. At the opening of the Niagara Falls Power Company, he declared,

> We have many a monument of past ages; we have the palaces and pyramids, the temples of the Greek and the cathedrals of Christendom. In them is exemplified the power of men, the greatness of nations, the love of art and religious devotion. But the monument at Niagara has something of its own, more in accord with our present thoughts and tendencies. It is a monument worthy of our scientific age, a true monument of enlightenment and of peace. It signifies the subjugation of natural forces to the service of man, the discontinuance of barbarous methods, the relieving of millions from want and suffering.

While Tesla certainly was a visionary, there were some things that he could not foresee. In the case of hydroelectricity, not all landscapes are blessed with spectacular waterfalls, so they must be made artificially with dams, which block crucial highways for marine animals who call the rivers home.

One of the biggest construction projects on Earth was China's Three Gorges Dam. Built across the Yangtze, the dam not only caused 1.3 million people to be displaced to make way for the 660-kilometre reservoir behind it, but it also had a devastating impact on the fish. A third of the nation's fish species once inhabited the river basin, but after the dam was built four species of carp suffered a 50 to 70 percent decline, and several other animal species were threatened with extinction, including the now functionally extinct baiji, or Yangtze River dolphin.

In delicate ecosystems like the Amazon and Mekong River basins, the same threats continue. And in countries like Canada, hydroelectric dams block spawning salmon from migrating upstream to reproduce. To combat this, hydroelectric companies have built structures ranging from the unnatural to the bizarre as workarounds. Fish ladders, essentially step pools going up the dam, allow fish to leap upward in their drive upstream, but in the attempt to move upstream many still get caught in turbine blades, up to 11 percent. Supersaturated water—essentially air bubbles—is another problem for the fish. As the water tumbles and churns below the dam, nitrogen gases concentrate into bubbles, and this dissolved gas builds up in the water where it's eventually taken in by the fish. That is, as they breathe, the gases enter their bloodstream. Gas bubble disease can disorient fish, but more seriously, if the fish pass through several dams and the concentrations build up to toxic levels, it frequently kills them.

"Fish cannons" are now being investigated as a potential forward pass above the dams. Like a giant pneumatic tube, a vacuum at the base of the dam sucks up the fish and pulls them over one hundred feet at thirty-five kilometres an hour until they pop out above the dam. The system, absurd as it sounds, is less traumatizing for the fish than being netted and then trucked or moved by helicopter, which is how some wildlife services transport the animals back to their spawning grounds.

But damming or diverting a river not only affects fish; it has an impact on people too. Neighbouring nations and multiple communities make their claims to rivers, but rivers don't heed human borders. As a result, by altering the flow of a river, those living upstream are interfering with a critical artery of both food and water.

WE ARE THE MOST POWERFUL SPECIES on Earth because we have devised extraordinary ways to harness power. One of the most controversial comes from a source that is invisible: we take the smallest units of ordinary matter—atoms—and split them into even tinier particles to create nuclear power. To do this, we use an element that fissions easily: uranium, specifically the isotope uranium-235. When uranium is bombarded with neutrons, its atoms split and a chain reaction of these splitting atoms creates a tremendous amount of heat. It's a high-tech way to make heat, but other than that, a nuclear power plant works in much the same way as most coal or gas power plants or even the power plant in Savoie that uses cheese. Like a kettle, it uses heat to boil water and create steam, and this steam turns the turbines which generate electricity.

Like all forms of energy production, however, nuclear can have some pretty hefty problems associated with it. The most feared of these is a meltdown.

When the Tohoku earthquake struck Japan in 2011, it formed a massive tsunami. The waves were so big and strong that even after travelling seventeen thousand kilometres to the coast of Chile, they were two metres high. Much closer to the epicentre, a mere 160 kilometres away, sat the Fukushima Daiichi power plant, which despite being constructed to withstand an earthquake and a 5.7-metre tsunami, would prove no match for the savage 15-metre-high waves. When the water came crashing through the walls, the plant's ground level fuel tanks for its generators were destroyed. Without power, the pumps built to circulate the coolant water shut down, causing the three reactors to overheat, which led to the meltdown.

It took six years to finally find the uranium fuel rods in Reactor 3. In the heart of the disaster zone, radiation levels in some spots were as high as 650 sieverts an hour—a person walking in would be killed in a minute. Instead, robots were sent in to search for the fuel rods, and even then multiple robots died on the job. Eventually, a shoebox-sized robot called Little Sunfish was able to swim through the flooded maze of the reactor and locate the uranium that had melted through the floor.

Eleven percent of the world's electricity comes from nuclear power. And while it's an energy source with a bad rap, it should be stressed that it is usually safe. The problem is those rare "acts of God" or unforeseen events where things spiral out of control. Then, things do go horribly wrong. In Japan, ninety-seven thousand people have yet to return to their homes, and some likely never will. Cleaning up the mess of Fukushima will cost an estimated $188 billion, and the site will remain contaminated for at least the next thirty to forty years.

IN A COMPLETELY FAIR WORLD, oil companies would pay *us* to use gasoline. Gasoline is a toxic by-product of the crude oil distillation process. Some of the other things that come out of a barrel of oil are ink, crayons, bubble gum, dishwashing liquids, deodorant, eyeglasses, records, tires, ammonia, and heart valves. There's also asphalt, lubricating oils, paraffin wax, heating oil, tar, and other ingredients of industrial products, especially petrochemical feedstocks for plastics. Diesel is the fuel you want for running big trucks, trains, and heavy machinery, and jet fuel has an obvious use. It's hard to fly jets without it. Gasoline is a by-product of Kerosene, which replaced whale oil in the nineteenth century as fuel to light lamps. Until oil companies found a way to market it to us, they just dumped it in nearby rivers. That is, until they found a way to get us to pay for it. (If that sounds crazy, consider that oil companies still flare off natural gas at the wellhead—the same natural gas you use to heat your house.)

Today, we can't imagine living without it, as anyone who has awakened to the sound of a neighbour's two-stroke leaf blower on a Sunday morning can attest. From one perspective, our fleets of cars and motorcycles, our jet skis and fishing boats, lawn mowers and chainsaws may only be expensive devices for disposing of someone else's toxic waste. But from another, they signify the good life. Cars especially.

In the world we live in, gasoline makes many tasks faster and easier. That's because gasoline is an incredibly dense package of energy. In *The Upside of Down*, Thomas Homer-Dixon worked out the calorific value of crude oil and found that it is approximately twelve thousand watt hours per kilogram. "Three large tablespoons of oil contain about the same amount of energy as eight hours of human manual labour," he writes, "and when we fill our tanks with gas, we're pouring into the tank about two years of human manual labour."

Oil itself, then, is power in the truest sense of the word, and when you understand how much hard labour it frees us from, you can see why we've become so addicted to it. Every single day, the world uses over ninety million barrels of oil.[8] And every litre of that ancient substance has incredible power. While humans used to rely on raw muscle power or the power of domesticated animals to work the fields, today machines running on oil can do much of that work for us. Unlike green energy, it is highly portable, which is why we think of cars rather than generators when we think of oil. But nothing is easier than generating electricity from oil if you can afford it, like Saudi Arabia can.

Of course, gasoline and diesel freed not only humans from physical labour but also vast numbers of oxen and horses. In particular, the "horseless carriage," otherwise known as the car, came into prominence because of the internal combustion engine's ability to transmute this prehistoric energy into motion. A car engine works by igniting the fuel in a series of rapid explosions, or what engineers call deflagrations. If you're sitting in the parking lot with the engine on—say you've got a four-stroke, four cylinder engine that's idling at 750 rpm—that works out to 1,500 sparks of combustion a minute. When the fuel

8 By 2020, the number is expected to increase to approximately 100 million barrels per day.

ignites, the explosive force causes a piston to move up and down, transforming chemical energy into mechanical energy to power the car. Today, we still have a reminder of life before the advent of the gas powered engine. The term "horsepower" gives us a sense of how much horse-equivalent energy is made by our engines.

The acceleration of our petrol use, and the rapid increase in technology that led to a corresponding increase in horsepower, came from the military. In the First World War, the average American division used 4,000 horsepower. By the Second World War, the average division in the war effort used over a hundred times more gasoline, or 187,000 horsepower. Even today, the biggest user of oil is the military. The US military alone uses one hundred million barrels of oil every year.[9]

The power that comes from oil, then, is not only machine power but also state power. The planet's superpowers are, not uncoincidentally, the nations that have access to and use the most oil. Lack of access to oil has meant a lack of power, a hard lesson learned early on. During the First World War, it became clear to Winston Churchill that oil played an integral role in offensive strategy. You could paralyze an army by cutting off its oil supply, because a country without oil would have no source of energy to run its ships, tanks, and planes.

You simply cannot fight a modern war without oil. Refined oil is an "indispensable material for laying runways, making toluene (the chief component of TNT) for bombs, the manufacturing of synthetic rubber for tires . . . and that is not to mention the need for oil as a lubricant for guns and machinery." We speak of wars

---

9 Oil price spikes can have a huge impact on the military, costing billions of dollars for every $10 dollars that a barrel goes up. Because of this, the military is also spearheading the use of green and solar technologies.

being fought over oil as if the oil were only an end goal when in fact it is required for modern warfare in the first place.

A country without oil is a country that can quickly be defeated. This is why, for the architects of war, it became so important to secure oil states like Iran and Venezuela. Having their oil meant having a constant flow of energy.

Since 1973, up to 50 percent of all wars between states have been linked to oil, and much of the blood spilled in the twenty-first century has been in the Middle East. Which, from a geologist's point of view, brings up a curious question. Why is there such a spectacular abundance of oil—some 60 to 70 percent of the world's supply—in this particular region?

To uncover the answer, we have to see beyond the gas pump and make a deep dive into prehistory, back to a time when the world looked very different not only in terms of our planet's inhabitants but also its geography. If we journey back to the mid-Cretaceous, some 85 to 125 million years ago, the continents were much closer together than they are now; they were, however, just beginning to spread out, having fractured and split off from the supercontinents of Gondwana and Laurasia.

The land masses were beginning their slow march into the configurations that we recognize today. North America and Eurasia had begun to trudge northward, and South America, the Middle East, Africa, Australia, and Antarctica had begun migrating their way slowly south. And between the northern and southern continents, curving up just above the equator, was a vast and ancient ocean that has long since disappeared.

Called the Tethys Ocean, after the ancient Greek sea goddess, it existed in a time when Earth was truly a water world. Only 18 percent of our planet was dry land, and water levels on average were 170 metres higher than they are today. With greater volcanic and tectonic activity, this was also a greenhouse world. Volcanoes belched vast amounts of carbon dioxide out into the atmosphere, and, by the late Cretaceous, atmospheric CO_2 was approximately four to eighteen times our current levels, making the planet a lot hotter than it is now. There were no polar ice caps; the water instead was a temperate 10°C to 15°C, while the equatorial ocean temperature was 25°C to 30°C. The critical point, as geologist and oceanographer Dorrik Stow argues in *Vanished Ocean*, is that warm water holds less oxygen.[10] The lack of oxygen, coupled with more sluggish circulation of water due to the high temperatures, created a suffocating marine environment that around ninety-four million years ago led to what created the vast oil fields in the Middle East: a large oceanic anoxic event[11] that Stow calls the "Black Death." In this environment, anaerobic bacteria broke down dead plants and animals much more slowly as they rained down on the ocean floor. The organic matter only partially decayed, leaving carbon in the sediment. Buried under layers of mud and silt, and over millions of years, the dead plants and animals were compressed and heated up by the roaring furnace at the centre of Earth.

That's what oil is: dead stuff. Much of which comes from an extinction event.[12] The high-tech world we live in is fuelled directly by a prehistoric one. Each time you turn your car

10 If you take two open soda cans and leave one at room temperature and put the other in the fridge, the colder one will be fizzier, as it can hold more dissolved gas. The same holds true of ocean water. Colder water can "hang on" to oxygen, while warmer waters release it into the atmosphere.

11 Researchers say there were between two and seven large oceanic anoxic events in the mid-Cretaceous.

12 University of Alberta scientists have evidence to suggest that underwater volcanism may have been responsible for a mass extinction event 93 million years ago which led to the formation of major oil reserves.

engine on and rev it up, it's like a funeral pyre, igniting ancient chemical remains. And since, as we know, life is made of carbon, with each combustion the molecular remains of these dead organisms turn into ghosts in the sky: the spirits, in effect, of carbon dioxide.

The average tank of gas holds what used to be more than one thousand tons of ancient life. A jaw-dropping twenty-three metric tons of prehistoric life goes into every litre of gasoline. That's the equivalent of forty acres of biomass being pumped into the tank so you can drive your car or SUV. According to Jeff Dukes, an ecologist at the University of Utah, "Every *day* [emphasis mine], people are using the fossil fuel equivalent of all the plant matter that grows on land and in the oceans over the course of a whole year."[13]

Unlike oil, coal was formed primarily by ancient forests of the Carboniferous era around three hundred million years ago. On land, this was the age of giants. The skyscrapers then were tall fern-like trees that towered over forty-five metres high above a landscape covered in thick, rich vegetation. In this hot, humid jungle, buzzing with massive insects, the trees were notably different from today's. Their roots didn't reach very far into the ground, and when the trees fell over, their whole, massive trunks accumulated in the forest swamps. Microbes that could digest the trees' cellulose and lignin had not evolved yet, so instead of rotting, the trees stayed whole and their carbon remained inside them. Over time, as more and more trees accumulated on the forest floor, the wood compressed into peat, and over millions of years it became the coal we use today.

These ancient forests have warmed our homes, powered transportation, run our machines and factories, and brought us electricity. But burning coal releases a choking amount of pollution.

---

13 A gallon (4 litres) of gasoline contains thirty-one million calories.

It blackened the skies during the Industrial Revolution, just as it creates unbreathable smog in China and India today. The toxic transmutation of coal also contributes to climate change. That's because for every ton of coal burned, almost triple that amount of carbon dioxide is released into the atmosphere.

Because of this, coal is thankfully and finally being phased out. Countries around the world are shutting old coal power plants down, but coal still contributes to our daily power use far more than renewables do. On the grid, it's responsible for 30 percent of the electricity made.[14]

Oil on the other hand, and the petrol that we pump into our cars, came primarily from marine life. In our ancient oceans, the waters were buzzing with a fascinating array of microscopic creatures. Just as a teaspoon of seawater would reveal a colourful bonanza of life today, looking under a microscope at the ancient seas, you'd see tiny zooplankton, phytoplankton,[15] and algae, as oblivious to our existence as we are to theirs.

It's often asked if larger animals like dinosaurs got into the mix. While it is possible, it should be noted that a significant amount of today's oil was deposited long before dinosaurs walked the earth. Some of the oil fields that we tap today are up to six hundred million years old. So while a few molecules of dinosaur here and there may be fuelling your ride to the supermarket, relatively speaking, compared to the vast amount of tiny

14 Coal demand has declined in Europe and the US, but has been offset by demand in India and other Asian countries.

15 Sunlight causes the bloom of over 5.5 billion metric tons of phytoplankton every year. These single-celled protists, which capture the sun's energy, are the primary producers of the food chain, capturing the sun's energy. Their cumulative death over millions of years captured and stored this energy, essentially making oil what is is: a massive natural battery.

plants and animals that make up our oil, dinosaurs have made an insignificant contribution.

Instead, this prehistoric stew formed as a constant flow of marine organisms died and drifted down to the sea floor. Buried under layers of silt and mud, in low oxygen conditions, they did not decay but rather formed into a waxy substance called kerogen. Typically, geologists note that "oil forms from organic matter that is either 'cooked' deep within the earth for long periods of time at low temperatures, or 'cooked' for short periods of time at high temperatures." Over time, the kerogen molecules break apart into hydrogen and carbon atoms. The heavier liquid mixture cooked at 50°C to 100°C turns into oil, and the lighter mixtures that were cooked at higher temperatures of 150°C to 250°C bubble up into rocky chambers and turn into gas.

Together, oil, coal, and gas are nature's oldest battery, one that took anywhere from a million to an average of a hundred million years to charge. That we have power to run modern civilization is because microscopic organisms captured ancient sunlight through the process of photosynthesis, just as today the sun feeds plants that in turn are eaten, as "food batteries," to fuel animals. The only difference with fossil fuels is we don't eat this ancient sunlight ourselves; it is food for our machines.

Today, the Middle Eastern oil states derive their bounty from an accident of geology. They were situated on the perimeter of the Tethys Ocean as the fossil fuels formed. The region is still responsible for the majority of the world's oil—two-thirds—and a quarter of its supply of gas. The bulk of the fuel comes from prehistoric life, formed by conditions in the Cretaceous.

It should not be lost on you, dear reader, that we are beginning to see parallels in our own warming world. Right now, as you read this, whole shelves of ice at the poles are calving and crashing into the ocean. At the equator, the surface temperature of the ocean is like bathtub water at a very warm 30°C. Anoxic waters are also spreading, and while they are nowhere as severe as what existed during the Cretaceous,[16] scientists have already documented a 2 percent decrease in the oceans' oxygen levels over the last fifty years, and this silent creep of anoxic water[17] has grown by over 4.5 million square kilometres. To offer some perspective, that's the size of the European Union.

For most of us, what takes place underwater is out of sight and out of mind, but scientists who are researching oxygen-starved waters are deeply concerned. Along with local die-offs of bottom-dwellers like sea stars, crabs, and anemones, deep-sea fish like marlin and sailfish that often feed at depths of up to eight hundred metres have recently been tracked feeding in much shallower waters. Scientists studying sailfish off the coast of Central America found that they were no longer venturing deeper because of a huge sink of oxygen-depleted water. The fish were staying up in the shallows because if they ventured farther into the band below, they would suffocate.

While we tend to think of oxygen as important here on land, we forget that the dissolved gas is just as vital to sea life as it is to terrestrial life up above. Imagine if vast regions of our air saw a similar decrease in oxygen: it would choke out the life that surrounds us.

---

16 Geochemist Martin Fowler has suggested that the levels of anoxia in the Cretaceous would be more like what we see in the Dead Sea.

17 "Already naturally low in oxygen, these regions keep growing, spreading horizontally and vertically. Included are vast portions of the eastern Pacific, almost all of the Bay of Bengal, and an area of the Atlantic off West Africa as broad as the United States . . . The zone off West Africa that's as big as the continental United States has grown by 15 percent since 1960—and by 10 percent just since 1995. At 650 feet (200 metres) deep in the Pacific off southern California, oxygen has dropped 30 percent in some places in a quarter century."

HUMANS ARE THE ONE SPECIES on Earth with artificial superpowers. We have taken the power of the dead, the power of the sun, the power of the wind, the power of water, even the power of invisible atoms, and we have harnessed all of this energy and transformed it so we can control the world around us beyond our natural capabilities. And while we may grow up reading about Superman in the comic books, picturing his skills as extraordinary, human beings now have the same superpowers, except we access them with the flip of a switch. Everything Superman can do—flight, X-ray vision, super strength, speed, heat vision, freezing breath, and the super flare (an explosive, omnidirectional blast that obliterates anything within a half kilometre radius)—we can do, as long as we have sufficient energy and the right technology. We can fly around the planet. We can fly into outer space. We can see and hear what's happening in different parts of the world as it happens. To our cave-dwelling ancestors, we would seem to possess magic. We would appear as powerful as gods.

Most of us know little about where it really comes from, or how it works; the source of humanity's power is a blind spot. But our power is also our kryptonite. Because energy is so accessible—there with the push of a button or the turn of a key in the ignition—we are blind to how much of it we use. The energy we require to keep us alive is about two thousand calories a day, which works out to approximately ninety watts, or the equivalent of one light bulb of energy for our metabolism. But cumulatively, to power all of our modern "stuff," we require much, much more. As physicist Geoffrey West writes, "We now require homes, heating, lighting, automobiles, roads, airplanes, computers, and so on. Consequently, the amount of energy needed to support an average person living in the United States has risen to an astounding 11,000 watts. This *social* metabolic rate is equivalent to the entire needs of about a dozen elephants."

As a global society, these needs are even further amplified. Together, we use approximately 150 trillion kilowatt hours of power a year. The world's citizens number approximately 7.5 billion, but our annual energy use is enough to supply a population of 200 billion.

As a result, we are pumping a frightening amount of carbon dioxide into the atmosphere. And yet we do so little about it. Why? As M. Sanjayan, a senior scientist at Conservation International, explains, "Right now there is CO_2 pouring out of tailpipes, there is CO_2 pouring out of buildings, there's CO_2 pouring out of smoke stacks, but you can't see it. The fundamental cause of this problem is largely invisible to most of us."[18]

What we don't realize and don't physically see is that the underground reservoirs of fuel we are tapping contain five times more carbon than the invisible reservoir of carbon dioxide in the atmosphere. The carbon cycle, in which carbon naturally moves through the atmosphere, takes three years; it remains in plants on average for five years, in soils for thirty years, in oceans for three hundred years, and goes through the geochemical cycle once every 150 million years. But in a very fundamental way, we are messing with this natural cycle by artificially injecting

---

18 How huge is this blind spot? It is so all-encompassing that we may not even see the problem when we're driving around in it. Cars with low gas mileage are obviously part of the problem. But what is less obvious is that, depending on where you live, your hybrid or electric vehicle might not be any cleaner than a gas-guzzling SUV. About a third of our electricity comes from coal, which is about as dirty as fuels come. Meaning, in some places, even the solution is part of the problem. Think of the fossil fuels that go into mining and smelting, manufacturing, transporting, and assembling a wind turbine before it starts pumping "clean" electricity into the grid, and you begin to glimpse just how deeply wedded even the most ambitious plans for the future are implicated in the energy of the past.

carbon found deep underground into the atmosphere. Today, there is 45 percent more carbon dioxide in the air than there was before the Industrial Revolution. The last time this much CO_2 was in the air was over eight hundred thousand years ago.

Once we've used it, the energy for our superpowers seems to vanish into thin air. But that's the grand irony: countries will go to war to decide who *owns* the hydrocarbons and then countries gather to decide who *does not own* the carbon dioxide.

In the end, however, there may be a clear way to visualize the heat buildup from all the fossil fuels that are burned. According to climatologist James Hansen, the unseen rate at which our planet is warming is the equivalent of dropping four hundred thousand Hiroshima bombs every single day.

6

TRASH & TREASURE

We are fast becoming a plastic society.
Pretty soon, we will have more in common
with Ken & Barbie than with our natural environment.
—ANTHONY T. HINCKS

IN OUR MIND'S EYE, the grey, cratered landscape of the moon is untouched. Up there still are the iconic first human footprints, the American flag, and a plaque that reads, "Here men from the planet Earth first set foot upon the moon, July 1969, A.D. We came in peace for all mankind."

After five decades on the moon, however, the flag has begun to surrender to the elements. Bleached by harsh UV rays from the sun, the Stars and Stripes have disappeared and the nylon has faded to white. But the Americans didn't just plant one flag on the moon; they planted six. And space travellers have left a much heavier footprint than simple human tread marks. Littering the lunar surface are 181,000 kilograms of forgotten trash.

According to NASA, along with ninety-six bags of urine and vomit, there are old boots, towels, backpacks, and wet wipes. With no garbage cans at hand, the astronauts also littered the landing site with magazines, cameras, blankets, shovels. And after several international missions, there are now seventy spacecraft on the surface, including crashed orbiters and rovers.

Compared to Earth, the moon has a very thin atmosphere,[1] so it will take some time for the evidence of our visits to erode and disappear. Arizona State University scientist Mark Robinson suggests that with the impact of particle-sized micrometeorites hitting the garbage, the evidence of our brief stays on moon will break down and be gone in about ten to a hundred million years.

Viewed from the lunar surface, our own planet rises above the horizon and shines into the night like a blue moon. From a distance, it too looks pristine, but up close you would see a gleaming cloud of space junk orbiting Earth. Our planet has come to resemble Pig-Pen from the *Peanuts* comic strip. Right now, there's almost three thousand metric tons of space junk continuously circling us.

This wasn't always the case, of course. In the 1950s, Earth orbit was junk-free. It was not until March 17, 1958, that it acquired a permanent resident. Today, this dead satellite, the Vanguard 1, holds the title of the oldest piece of orbital debris. It completes a full revolution around Earth every 132.7 minutes. But it is no longer alone. It's been joined by more than 29,000

---

1 Contrary to conventional wisdom, the moon does have an atmosphere, though it is insignificant compared to the density of Earth's atmosphere. The technical term for this type of collision-free atmosphere is a "surface boundary exosphere."

other pieces of space junk invisibly circling us, along with over 1,700 active satellites. The U.S. Air Force has been tracking orbital debris, which is mostly made up of spent rocket stages and decommissioned satellites, and keeps a record of any object larger than a baseball. Parts do break loose that are smaller. Everything from paint chips, nuts, bolts, bits of foil, and lens caps are among the 670,000 objects that are one to ten centimetres in size.

As the size of the objects decreases, the number of them increases. For debris that ranges from a millimetre to a centimetre in size, the number is approximately 170 million. But just because they are small doesn't mean they are harmless. According to the European Space Agency, a one-centimetre object moving at orbital speed could penetrate the International Space Station's shields or disable a spacecraft. The impact would have the energy equivalent of an exploding hand grenade.

But we don't only dump our spacecraft in space. We also dump them in the sea. In the Pacific Ocean, miles under the waves, is a site called Point Nemo, which serves as a spacecraft cemetery. Chosen for its remoteness (the closest land mass is nearly 2,400 kilometres away), it is where international space agencies discard large space objects that don't burn up in the atmosphere upon re-entry. From 1971 to 2016, over 260 spacecraft were dumped at Point Nemo. The junkyard became the final destination for 140 Russian resupply vehicles, a SpaceX rocket, the Soviet-era Mir space station, and several of the European Space Agency's cargo ships, all of which lie on the ocean floor, slowly disintegrating.

At launch, we marvel at these multi-billion-dollar technological masterpieces, but once they've outlived their use, like all objects, no matter how advanced or expensive, they become garbage. Humans are a tool-making species, but as a consequence we are also a trash-making species. And while we don't have a love-hate relationship with our things, we do have a "love-indifference" relationship with them. We covet objects before we own them and later throw them away without thinking about them again. That's the thing about our garbage: we have become experts at acting like it doesn't exist. Space trash, in fact, barely registers as a blip compared to the enormity of the waste our species generates. In disused home appliances, computers, mobile phones, and other electronic equipment, or e-waste, we generate close to forty-five million metric tons of waste every single year. That's the equivalent of over 4,500 Eiffel Towers. Trash that could obstruct a city skyline. But not only do we not see it, most of us don't even know where it goes.

There are some things we do know about our trash. The world leader in trash production, for instance, is the United States. Around the world, rich countries and rich people produce more garbage. Individually, each American throws out about 3.2 kilograms of garbage a day, or over ninety metric tons of garbage in a lifetime. As Edward Humes writes in *Garbology*, "A single person's 102-ton [US] trash legacy will require the equivalent of 1,100 graves. Much of that refuse will outlast any grave marker, pharaoh's pyramid or modern skyscraper."

But even then, what we toss out is just the tip of the proverbial trashberg. Most garbage comes from the manufacturing process. What we throw in the bin—the final product—represents a mere 5 percent of the raw materials from the manufacturing, packaging, and transportation process. Put another way, for every 150 kilograms of product we see on the shelves, behind the scenes there's another 3,000 kilograms of waste that we *don't* see. In total, the world produces approximately three million metric tons of garbage every twenty-four hours. That number is expected to double by 2025. And if business continues as usual, by the end of the century it will be an unfathomable ten million metric tons of solid waste a day.

It isn't just our factories that create waste. As biological beings, we generate our own waste as well. And with 7.5 billion people on the planet, that crap adds up. In *The Origin of Feces*, David Waltner-Toews charts the meteoric rise of human excrement: "In 10,000 BCE there were about a million people on the planet. That's 55 million kilograms of human excrement scattered around the globe in small piles, slowly feeding the grass and fruit trees . . . By 2013, with more than 7 billion people on Earth, the total human output was close to 400 million metric tons (400 billion kilograms) of shit per year."

With such colossal amounts of human biological waste and manufactured solid waste, it's like a magic trick of epic proportions that it all just seems to—*poof*—disappear.

Before the days of the garbage collector, though, people had to deal with their shit, literally. There was no getting away from it, because it sat, steaming, fly-ridden, and reeking right in front of us. The familiar Brooklyn stoop we all know from Sesame Street is not just an architectural carry-over from the Dutch, it was also a way of dealing with nineteenth-century waste. The steps lead up to the parlour floor because at the time in New York, people threw their garbage out of the windows and right on to the city streets. The trash was so high—up to a metre in the winter when it combined with snow and horse waste (the latter of which piled up at a rate of 1,000 metric tons of manure and 227,000 litres of urine every day)—that the stoop allowed people to get up above the mess and make their way safely in the front door.

Nineteenth-century waste management was assisted by scavenging dogs, rats, and roaches, but the primary street cleaners were pigs. In the United States, piggeries were specifically erected for big towns with populations of over ten thousand. Our trash was their dinner, with an average of one metric ton of waste digested by seventy-five pigs a day. It's not uncommon to find paintings of New York City at the time featuring these roaming pigs. For the Europeans who painted them, the urban swine were a novelty, but for New Yorkers the fact that hogs ran wild in the streets was pretty much standard fare.

Up until the 1840s, thousands of pigs roamed around Wall Street. Today, the area is known for its bankers and high-rollers, but the name Wall Street, from the original Dutch "de Waal Straat," derives from a 3.5-metre fence built to keep hogs from causing damage to the streets and residents' gardens.

In Paris, trash and human waste also flooded city streets. The French were the first to establish a corp of sanitation workers and began managing city waste in this manner four centuries earlier. But streetside filth was a continual problem, leading the French king to hand out an edict to deal with the squalor in 1539:

> François, King of France by the Grace of God, makes known to all present and all to come our displeasure at the considerable deterioration visited upon our good city of Paris and its surroundings, which has had in a great many places so degenerated into ruin and destruction that one cannot journey through it either by carriage or on horseback without meeting with great peril and inconvenience. This city an its surroundings have long endured this sorry state. Furthermore, it is so filthy and glutted with mud, animal excrement, rubble and other offals that one and all have seen fit to leave heaped before their doors, against all reason as well as against the ordinances of our predecessors, that it provokes great horror and greater displeasure in all valiant persons of substance.

In Paris, waste became a private affair. Instead of putting it out in the streets, Parisians were ordered to build cesspools in their backyards. Inevitably, the neighbourhood stench, along with bouts of cholera, became far too much to bear.[2] The French switched over to a method the Chinese had been using for thousands of years: managing their population's waste by turning it into "night soil," a euphemism for human excrement used as manure for farming.

By the 1800s, what growing cities had discovered was that a city, by its very nature, localizes and concentrates waste on a massive scale. They become, for lack of a better term, engines for producing giant shit heaps. The Chinese had been diffusing the situation by taking their excrement from populous areas and returning it to the countryside. There, it wasn't waste. It was brown gold. Human manure was returned to the soil in order to feed the nation. The system, in fact, worked very well, and until recently China was renowned for its fertile soil and sustainable agriculture. For thousands of years, about 90 percent of human manure was recycled in China and accounted for a third of the country's fertilizer.

Consider for a moment your own digestive contribution. On average, you excrete about fifty to fifty-five kilograms of feces and about five hundred litres of urine a year. But this "waste" contains valuable nutrients. According to the German Corporation for International Cooperation, on an annual basis that works out to about "10kg of nitrogen, phosphorus and potassium compounds, the three main nutrients plants need to grow—and, helpfully, in roughly the right proportions." One person's excrement is enough to fertilize and grow over two hundred kilograms of cereals a year.

The Japanese also recognized the value of shit. During the Edo period (1603 to 1868), in the area that is now Tokyo, the Japanese ran a closed-loop system, and *shimogoe* (translated as "fertilizer from the bottom of a person") became critical for sustainable

---

2 In 1832, 20,000 people died of cholera in Paris alone.

agriculture. On roadsides near the fields, buckets were provided for travellers, who were encouraged to leave their waste behind. As David Waltner-Toews writes, "The seventeenth-century city of Edo sent boatloads of vegetables and other farm produce to Osaka to be exchanged for the city's human excrement. As the cities and markets grew (Edo had a million people by 1721) and as intensive paddy-farming increased, prices of fertilizers, including night soil, rose dramatically; by the mid-eighteenth century, the shit owners wanted silver—not just vegetables—for payment."

Crap had become a high-priced commodity. Landlords could increase the rent they charged if the number of tenants dropped in their building, because with fewer defecators to pad an owner's income, running the property became less profitable. As a business, managed through private agents and not the government, *shimogoe* prices were set by the landlords, leading to conflict with farmers, who were often gouged with high prices.

There was also good shit and bad shit. Rich shit surely stank as much, but it was more highly prized. As the rich ate more diverse diets, this resulted, according to the farmers, in better nutrients in their feces.[3] As for its value, the price of *shimogoe* depended on demand, but at its height rose up to 145 mon per household. For perspective, in 1805, 100 copper mon could buy a good lunch of mushrooms, pickles, rice, and soup. By the 1800s,

---

3 "Human night soil is essentially the residue of what people eat after they have absorbed necessary nutrients. The night soil of a population that ate a lot of fish and meat generally contained more nitrogen and phosphate. People whose diet was mostly vegetarian (cereals and vegetables) had night soil that was generally poor in nitrogen and phosphates, but rich in potassium and salt."

—Tajima, Kayo. 'The Marketing of Urban Human Waste in the Early Modern Edo/Tokyo Metropolitan Area'. Environnement urbain : cartographie d'un concept. Vol 1. (2007).

the price of human waste was so valuable that stealing it became a criminal act that could result in imprisonment.

Human waste was also ranked in comparison with compost and other animal manure. In an 1849 issue of the American magazine *Working Farmer*, the eminent German agriculturalist Professor Hembstadt is quoted as saying,

> If a given quantity of land sown without manure, yields three times the seed employed, then the same quantity of land will produce:
>
> Five times the quantity sown when manured with old herbage, putrid grass or leaves, garden stuff, etc. etc.,
> Seven times with cow dung,
> Nine times with pigeon dung,
> Ten times with horse dung,
> Twelve times with goats' dung,
> Twelve times with sheeps' dung, and
> Fourteen times with human urine or bullocks' blood.

But for those steeped in the fine art of stercoration, there was one type of excrement that was always top of the list. When it came to the best fertilizer in the world, there no competing with guano.

People have gone to war over many things in history, but the Guano War of 1864 to 1866 may have been the first time a war began over the sovereignty of bird shit.[4] The guano was a virtual goldmine for Peru. Spain knew this and was determined to reassert its power and seize it from its former colony. As a result, Chile joined the two-year war, and the South American countries fought together to fend off their former colonizers.

Coming in by boat, you smell the Chincha Islands long before

4 The Guano War is also known as the Chincha Islands War.

you see them. With nesting colonies of pelicans, boobies, and cormorants, the Peruvian archipelago was home to over a million birds. Each bird produced about twenty precious grams of droppings a day, together producing around eleven thousand metric tons per year. Over generations, and with little rainfall in the area, the mounds grew into mountains. And by the early 1800s, the guano on the Chincha Islands was over ten storeys tall.

Guano's property as a powerful fertilizer had been known to the locals for centuries; they called it *huanu*. Seabird excrement is particularly potent because it's packed with marine nitrogen. As the birds feed on huge schools of anchoveta and plankton, they act as "biological pumps" that transfer the nitrogen into terrestrial ecosystems.[5] This gift of soil fertility was so highly valued that for the Incas killing a seabird could result in a death sentence.

The Europeans came to realize its value when explorer Alexander von Humboldt first brought some back with him in 1804. For the farmers using it on their land for the first time, the results seemed miraculous. Exhausted soils suddenly became fertile again, and it boosted crop yields by 30 percent. Unlike regular barnyard manure, guano was special shit: according to one expert, it was thirty-five times more powerful.

By 1850, as science writer Thomas Hager notes, the Chinchas—these barren islands covered in bird shit—were "acre for acre . . . the most valuable real estate on earth." A "guano mania" had taken hold. Tens of thousands of metric tons of guano were exported every year, accounting for up to 60 percent of the Peruvian economy. The Americans, eager to secure their own sources of guano, passed the Guano Islands Act on August 18,

---

5 Bird poop brings 3.8 million metric tons of nitrogen out of the sea each year. The nitrogen comes from dissolved gases in the air that mix with the water, and are broken into fixed nitrogen. During the 1800s this process was largely done by cyanobacteria.

1856, essentially allowing the United States to lay claim to any island they found with guano deposits. As stated in Section 1 of the act: "Whenever any citizen of the United States discovers a deposit of guano on any island, rock, or key, not within the lawful jurisdiction of any other Government, and not occupied by the citizens of any other Government, and takes peaceable possession thereof, and occupies the same, such island, rock, or key may, at the discretion of the President, be considered as appertaining to the United States." To date, over a hundred islands in the Pacific and Caribbean have been claimed, and while most titles were relinquished after the guano was exhausted, the act is still in effect today.

Eventually, that became the problem with the Chinchas. The guano was a finite resource that could not be replenished as quickly as it was extracted. By the time of the guano war (which Spain lost to the united front of Chile and Peru), there was less than a decade's worth of guano left. When it was gone, Peru went bankrupt.

One man saw the disaster looming and realized that Europe would soon be in very deep shit, figuratively speaking. With the primary guano source depleted, the fertilizer business had moved on to Chilean nitrates, a white granular substance found in the desert that was the next best thing. But William Crookes, an English scientist, had run the calculations. By his estimate, at the current rate of demand even the nitrates would be gone within decades. In his presidential address before the British Association for the Advancement of Science in 1898, the esteemed chemist sent out the clarion call before a packed house: "England and all civilised nations stand in deadly peril of not having enough to eat. As mouths multiply, food resources dwindle. . . . I hope to point a way out of the colossal dilemma. It is the chemist who must come to the rescue of the threatened communities. It is through the laboratory that starvation may ultimately be turned into plenty. . . . The fixation of atmospheric nitrogen is one of the great discoveries, awaiting the genius of chemists."

What Crookes was urgently calling for was the development of synthetic manure. But despite his prophetic remarks, the world had no way of knowing that this fertilizer would literally come from thin air.

IT HAS BEEN CALLED the greatest invention no one has ever heard of. Without the Haber-Bosch process, half the people on the planet would not be alive today. It was developed in answer to Crookes' rallying cry to the chemists as a way to feed the world without relying on the two primary sources of fertilizer at the time: the now all but dwindled supplies of Peruvian bird poop and the strategically held reserves of Chilean desert nitrates.[6]

What both of the earlier sources had in common was that they were rich in fixed nitrogen. And while nitrogen is plentiful in the air around us—it makes up 78 percent of what we breathe—the kind of nitrogen plants need to take up from the soil comes in a different form, as fixed nitrogen. On land, it's made naturally in one of two ways. The first and most dramatic is through lightning. During storms, bolts of electricity are powerful enough to crack apart the tight bonds of atmospheric nitrogen, and when it comes into contact with water the element takes the form of nitric acid, which then leaches into the

---

6 Nitrates had a dual purpose, they could be turned into fertilizer or explosives. For Europeans, it took three whole months for a shipment to arrive. The Germans, in particular, were keen to create their own source of nitrates since they knew that during war time, they could be blockaded, which would cripple their ability to grow food and keep them from replenishing supplies of gunpowder.

soil. The second is from types of bacteria that have formed a symbiotic relationship with certain beans and legumes. Using a complex set of enzymes, these bacteria are able to break the nitrogen bonds, making it available to plants at their roots.[7]

Nitrogen in the air is considered "unusable" because the molecule N_2 consists of two super-tightly bound nitrogen atoms, one of the strongest bonds in nature. The atoms are locked together so securely, it takes an immense amount of energy—in the order of 1000°C—to tear them apart. So while we can breathe in and exhale atmospheric nitrogen, in this form it is also inert and cannot be absorbed into our bodies. Instead, the nitrogen that makes up our blood, skin, and hair comes from the food we eat. And it is essential. Nitrogen is found in every gene and every protein in living things. We couldn't exist without it, because it serves as the atomic backbone of our DNA.

The genius of the Haber-Bosch process was that it could "mine" nitrogen right from the air. Named after Fritz Haber, the scientist who invented it, and Carl Bosch, the engineer who industrialized it, the invention promised the world unlimited fertilizer. Finally, a source had been found that would not run out, because atmospheric nitrogen was everywhere. But while this "synthetic manure" relied on clever chemistry, it was not very simple to make. To scale the process up for mass production meant the Germans now faced another huge challenge: they had to build the world's largest machine.

Covering almost eight square kilometres, the factory they

---

7 Today, approximately 90 million to 120 million metric tons of nitrogen in our food system comes from natural processes, such as nitrogen-fixing bacteria and lightning strikes.

used, in Leuna, Germany, was "the size of a small city."[8] It housed massive compressors able to subject gases to two hundred atmospheres of pressure, about the same amount of pressure, as Thomas Hager writes in *The Alchemy of Air*, necessary to "crush a modern submarine." The process itself is not too complex: nitrogen and hydrogen gases are heated to a high temperature and then circulated over an iron catalyst,[9] which lowers the energy threshold of the reaction. The gas mixture is then put under so much pressure and heat that the hydrogen and nitrogen atoms crack and form a new bond, coming out of the machine on the other side as liquefied ammonia, or NH_3. Taking nitrogen from the air, Haber and Bosch had created a whole new way to feed plants. As the Germans said, it was *Brot aus Luft*. They were getting "bread from air."

Today, factories all around the world use the Haber-Bosch process to make synthetic nitrogen fertilizer. In 2016, they produced 146 million metric tons. And as the human population grows, the demand rises. In fact, the production of synthetic nitrogen and the rise in population are intimately linked. If you've ever wondered how the human population jumped in a single century from 1.6 billion people in 1900 to over 7.6 billion today, it is because we no longer use manure to grow food. This

---

8 It was twice the size of the first pilot plant, at Oppau, which exploded, killing six hundred people and injuring two thousand after the fertilizer in a storage silo caked together. Mixed in with sodium nitrate that was being manufactured for gunpowder, the combination had proved unstable, triggering a blast that is still recognized as one of the worst industrial accidents in history. Today, the Leuna works covers thirteen square kilometres.

9 After much testing the team settled on iron with aluminum oxide and calcium as the catalyst.

form of fixed nitrogen, in combination with the development of pesticides and new crop varieties, brought in what's known as the green revolution. Humans had tamed the earth, and their numbers exploded as a result. We could feed ourselves in a whole new way, by turning air into food with the use of synthetic fertilizer.

But there is one more *Matrix* pin-drop before we move on. Because half of the nitrogen in our food chain is now synthetically made, half of the nitrogen in *your DNA* comes from a Haber-Bosch factory.

EVERY YEAR, EIGHTY-THREE MILLION PEOPLE are added to the population of the planet. More people means more waste. And since the development of the Haber-Bosch process, a scandalous proportion of that waste has been uneaten food. To put it in perspective, the United States produced more than thirty-one million metric tons of food waste in 2010. According to the United States Environmental Protection Agency, by weight that is ten times more food waste than e-waste that year.

All of the energy used to grow, ship, and sell food that is ultimately tossed away is also wasted. In the United States in greenhouse emissions alone, it works out to all of the offshore oil and gas reserves being drilled for nothing.[10] On a global scale, according to the UN's Food and Agriculture Organization, approximately one-third of human food production does not get eaten. That is a jaw-dropping 1.3 *billion* metric tons of food that gets thrown out every year.

On top of *that*, there's another aspect to food waste. And it

---

10 All of that excess carbon dioxide may be invisible to us, but it's still getting pumped out there. Food waste alone accounts for 3.3 billion metric tons of CO_2 annually. That's more than two and half times the CO_2 emissions of every vehicle in the United States combined.

takes the form of the synthetic fertilizer made by the Haber-Bosch process. We use an enormous amount of fertilizer: for every person on the planet, there's approximately twenty kilograms of ammonia spread annually onto the fields. But a mere 15 percent of that manufactured nitrogen makes it into our mouths in the form of food[11]; the vast majority of our chemical fertilizers dissolve into waste.

As rains fall over the land each spring, nitrogen and phosphorus compounds from fertilizers are carried off into streams, rivers, and lakes and eventually drain into the ocean.[12] Here, the mixture of nutrients from our fertilizer runoff and sewage sparks an algal feeding frenzy, causing algae to spread over tens or even hundreds of square kilometres. Inadvertently, we are fertilizing the ocean. The resultant "bloom," however, is deadly. Marine plants and animals living beneath the thick, slimy mat of algae are deprived of sunlight. And when the algae overgrowth dies and sinks to the ocean floor, the massive decomposition process robs huge amounts of oxygen from the water, making it impossible for marine life to breathe. Those species that can't move to another location will not survive, and the biological desert that's left is known as a dead zone.

There are over five hundred of these dead zones in the oceans, and they are growing bigger. The fertilizer that was supposed

---

11 "Worldwide about 80 percent of nitrogen harvested in crops and grass goes to feed livestock instead of feeding people directly. Much of that nitrogen winds up in their manure and then gases off as it sits in giant open lagoons near intensive animal production centers or when it is spread onto fields without being properly mixed into the soil." Mingle, Jonathan. 'A Dangerous Fixation'. *Slate.* (2013, March 12).

12 You can have too much of a good thing. The process is a like overfeeding fish. We've artificially added twice as much nitrogen and three times as much phosphorous as would occur in a natural system.

to make life flourish is turning coastlines into graveyards. By disrupting the balance of nature with our human-made systems of survival, we have created a vicious cycle: now we need more fossil fuel energy (the equivalent of about 2.5 metric tons of TNT per acre) to create more food, in turn creating more mouths to feed. And every year the cycle escalates.

The Haber-Bosch process alone uses up almost 2 percent of the world's energy supply. And for every metric ton of ammonia created, two metric tons of carbon dioxide are released into the atmosphere. We are blind to this nitrogen waste in the ocean, just as we are blind to the carbon dioxide waste we can't see. But there is one form of waste we *can* see when it gets out of hand: air pollution.

IN BEIJING, A NEW COLOUR was named by the general public in November 2014. They called it "APEC blue." It was the result of a mission that began months earlier, when the Chinese central government tasked 434,000 staff in the regions of Beijing, Shandong, Tianjin, Shanxi, Hebei, Inner Mongolia, and Henan with orders to execute a grand plan. The teams had one ambitious goal: to change the colour of the sky.

In the days leading up to the arrival of international delegates for that year's Asia-Pacific Economic Cooperation (APEC) summit, 11.4 million vehicles were ordered off the roads and over ten thousand industrial plants suspended production. Under strict supervision, close to forty thousand other factories were put on a rolling schedule to limit their working hours and consequently the smoke and exhaust fumes they normally emitted.

The plan worked spectacularly. For two weeks in November, Beijing's notoriously thick grey-brown fog cleared and the air pollution fell by a jaw-dropping 80 percent. In its place, ready to welcome foreign dignitaries, leaders, and the world press, were soft white clouds and a brilliant APEC blue sky. But soon after the summit ended, the blue was gone too.

Today, Chinese citizens dream wistfully of the APEC blue skies of 2014, or the "military parade blue" skies of 2015. Scientists, for their part, have worked out why the skies darken so quickly once the temporary constraints on polluting are removed. It turns out that after the quick fix for a special event is over, there's an industrial backlash. Following the sharp drop in pollution for the duration of the event, there is a "retaliatory spike" of polluting as businesses ramp up to compensate for lost time and money.[13] Not surprisingly, perhaps, there is a direct relationship between economic activity and pollution.

In polluted cities, the term "AQI" is as familiar to anyone as "Celsius" or "Fahrenheit." It stands for air quality index, a scale designed to go from 0 to 500. Just by looking at the degree of haze, experienced residents can calculate the air quality. If you see a bit of haze on the horizon, that's 100. By 200, the grey horizon has closed in on you. At 300, the pollution haze is blocking the sun.

An AQI of 300 and above is considered hazardous to human health. Health effects include "serious aggravation of heart or lung disease and premature mortality in persons with cardiopulmonary disease and the elderly; serious risk of respiratory effects in the general population." Off the charts, at over 700, the air is described as industrial smoke. It's so thick that it's "chemical-tasting, eye watering." Coupled with a sandstorm, on May 4, 2017, the air was literally breath-taking. With an AQI of 905, Beijing had gone three times past the hazardous limit.

On bad days, let alone severe days, spending even twenty minutes outside can leave people feeling sick. Sore throats and

---

13 Political blue sky days are 4.8 per cent lower than average levels, but readings in the 4 days afterwards are 8.2 per cent higher.

coughs without cold or flu symptoms have become common. And for residents, especially those who live and work near factories, the coughs do not seem to go away.

The face masks worn by the Chinese public are now iconic. But in Beijing, only one section of society has it relatively easy when the pollution gets out of control. The rich can afford to protect their well-being by insulating themselves from the choking skies.

In the capital, the wealthy can send their kids to private schools, many of which have giant "playground bubbles" for children to play in. These pressurized air domes are equipped with hospital-grade air filters to purify the air and provide perfect "weather" year-round. On days when the AQI requires the children to stay inside, they are kept safe behind air-locked doors.

This pollution-proofing doesn't come cheap. Air domes cost millions of dollars, and even at home it costs tens of thousands of dollars to maintain fresh pockets of air for families to breathe in. Apartments in luxury high-rises are equipped with the latest high-tech air and water purifiers to provide a semblance of normalcy.

The poor have no option but to breathe the bad air around them. And it's not just China: India is home to 11 out of the 12 worst polluted cities. Likewise, Saudi Arabia and Iran have cities with pollution levels that make them hazardous to live in. According to the World Health Organization (WHO), which monitors a database of 3,000 cities in 103 countries, over 98 percent of cities in low- and middle-income countries failed to meet WHO air quality guidelines, whereas in high-income countries the failure rate nearly halved to 56 percent.

Of course, our bodies have inbuilt biological air filters—our lungs—and examining them can reveal the particulate matter we absorb from the outdoors. Pathologist Paulo Saldiva, who serves as a member of the scientific committee of Harvard University's School of Public Health and as a member of the air quality committee of the WHO, has done autopsies on the lungs of people exposed to outdoor air pollution. Blackened and pock-marked with carbon, they could easily be mistaken for the charred lungs of a cigarette smoker.

Every day, we inhale about twenty-three thousand breaths, taking in, on average, twelve thousand litres of air. The tiny hairs in our noses and the cilia that protect our lungs filter out the larger particles, but the most dangerous particles are the small ones, called $PM_{2.5}$, because the particulate matter is less than 2.5 micrometres in size. Collectively, they are the invisible sandstorm of sulphate, nitrates, black carbon, mineral dust, sodium chloride, and ammonia that we call "pollution."

Originating from the exhaust of car engines, mines, power plants, and industrial boilers, these incinerated particles have been strongly linked to lung cancer, kidney and cardiovascular disease, as well as asthma. In China, already the country with the highest rate of lung cancer, medical experts expect the number of lung cancer patients to rise to over eight hundred thousand a year by 2020. This is a silent epidemic. Worldwide, the WHO estimates that three million people die prematurely from outdoor air pollution every single year. By comparison, the number of people who die from AIDS is about a third that number, or 940,000.

When you sit back to consider it, the pollution we produce on an annual basis is staggering. It includes:

MANUFACTURED CHEMICALS: 30 million metric tons a year
PLASTIC POLLUTION OF OCEANS: 8 million metric tons a year
HAZARDOUS WASTE: 400 million metric tons a year
COAL, OIL, GAS: 15 gigatonnes (billion metric tons) a year
METALS AND MATERIALS: 75 gigatonnes a year
MINING AND MINERAL WASTES: roughly 200 gigatonnes a year
POLLUTED WATER (MOSTLY CONTAMINATED WITH ABOVE WASTES): 9 trillion metric tons a year.

These are the ingredients for a massive toxic bomb, according to veteran science journalist Julian Cribb. Globally, we build one of these bombs every single year. The difference is there's no deafening boom. Instead, it's a quiet fallout: the invisible particles seep into the food we eat, the water we drink, and the air we breathe. As Cribb writes, "Industrial toxins are now routinely found in newborn babies, in mother's milk, in the food chain, in domestic drinking water worldwide. They have been detected from the peak of Mt Everest (where the snow is so polluted it doesn't meet drinking water standards) to the depths of the oceans, from the hearts of our cities to the remotest islands. . . . The mercury found in the fish we eat, and in polar bears in the Arctic, is fallout from the burning of coal and increases every year."

The idea that there is some sort of "outside" world we can exist independently from is an illusion. Science shows us what our own eyes can't see: that everything in existence is part of a network, part of a flow. What we put out into the environment will eventually find its way back into our bodies.

OVER THE LAST THREE DECADES, hundreds of new cities have popped up in China. Today, the country has over six hundred cities, most of which were villages or small towns until very recently.

China's rise to economic dominance is a result of its manufacturing sector. The Chinese made the world's stuff, and they made it cheap. The colossal energy used to fuel and feed its population, manufacture goods for export, and build its new cities unsurprisingly now means that China is responsible for the highest amount of carbon pollution. By the end of 2017, China was responsible for 28 percent of global emissions, a significant portion of the estimated forty-one billion metric tons of carbon dioxide the world emits every year.

If we could see all those forty-one billion tons of CO_2, it would be like looking at the equivalent in tonnage of forty-one Mount Everests.[14] The fact that we can't has become the greatest challenge in any discussion of climate change. But there is another, more visible, way to see the effects of fossil fuel waste. In this form, it is everywhere around us: I'm talking about plastic.

As the American Chemistry Council notes, "Most plastics are based on the carbon atom. . . . The carbon atom can link to other atoms with up to four chemical bonds. When all of the bonds are to other carbon atoms, diamonds or graphite or carbon black soot may result. For plastics the carbon atoms are also connected to . . . hydrogen, oxygen, nitrogen, chlorine, or sulfur." It's shocking when you think about it, but just one hundred years ago plastic didn't even exist. As Edward Humes writes in *Garbology*, "Plastic has gone so fast from zero to omnipresent that it's slipped beneath conscious perception. Take a moment to scan the room you're sitting in. Everything from pill bottles . . . to the knobs on kitchen cupboards to the buttons on your pants to the elastic in your socks to the foam inside your seat cushion to the bowl you put your dog's dinner in to the composite fillings in your teeth . . . it's everywhere."

Today, the average person in North America uses one hundred kilograms of plastic every year, most of which is in the form of packaging that gets thrown out. When it was first invented however, plastic was durable and created to last. In 1907, the chemist Leo Baekeland developed it as a chemical replacement for an East Asian insect called the lac beetle. The arboreal insects secreted a hard substance called shellac that was harvested by manually scrapping it off the bark of trees. In the beginning, it was the material that the nascent electrical industry was using to insulate their wires.

---

14 The weight of Mount Everest is approximately 1 billion tons.

Baekeland thought there had to be a better way, so he set his mind to developing a synthetic substitute. In his lab, using a mixture of formaldehyde and phenol (an acid derived from coal tar), he produced a thick and sticky resin. On its own, it wasn't particularly useful, but when he added filler like wood dust or asbestos to the mixture, it gave the substance a surprising strength. Better yet, when he injected it into moulds, he found that he could shape it. He had created the world's first thermosetting plastic. Baekeland had created synthetic shellac.

Shiny new objects made from the novel material, called Bakelite, were first introduced to America in 1927. It was seen as a miracle substance. Now, instead of relying on elephant ivory for cutlery handles or tortoiseshell for glasses frames, there was an alternative, a plastic that could be shaped into anything. By 1944, Bakelite was being used in fifteen thousand products.

Around this time, mass production of the everyday plastics we use today began to take off. PVC, superglue, Velcro, Lycra, polyethylene bags, and polystyrene foam all came onto the market in the 1940s and 1950s. But still, global plastic production was less than a million metric tons per year.

Since that time, we have made over eight billion metric tons of plastic. Six billion metric tons of which we trashed.

In the United States, only a small amount of petroleum, approximately 5 percent, goes into plastic production. This brings a stark realization: all the plastic around us represents only a tiny proportion of the hydrocarbons in the environment.

We see our everyday objects but we are blind to the processes behind them. Looking at a plastic shampoo bottle for example, we don't see the oil spills, the landfills, or the Great Pacific Garbage Patch. As billions of people do everyday, we use a plastic item once—an object that will last ten thousand years—then throw it in the bin. The absurdity, as a popular internet meme has described it, is that "our society has reached a point where the effort necessary to extract oil from the ground, ship it to a refinery, turn it into plastic, shape it appropriately, truck it to a store, buy it, and bring it home is considered to be less effort than what it takes to just wash the spoon when you're done with it."

This was not always the case. Until the 1950s, household objects were valued. People cherished quality. Over generations, they passed on silverware or perfume bottles or chairs or dining tables or bed frames. They became our antiques. But in 1955, a new kind of lifestyle began to emerge. An August 1955 issue of *Life* magazine contained a vision of the new American family. The article title was "Throwaway Living" and the accompanying photo featured a man, woman, and child joyfully tossing disposable household objects in the air like confetti. Material prosperity meant that products were now being designed and promoted to be short-lived. As *Life* magazine proclaimed, "disposable items cut down household chores."

Soon, along with single-use plastics, products had "seasons," and consumers were expected to keep up with "trends." Planned obsolescence became a part of the design principle.[15] If products did not go out of style, then they were designed to fail after a certain period so that new products could be bought to replace them. It's important to remember that, relatively speaking, this is an extraordinarily recent shift in humanity's way of thinking. It has, however, had a monstrous effect. Today, despite nationwide calls to ban plastic, along with the creation of new forms of

---

15 The most famous example of planned obsolesce is the light bulb. Today the average incandescent lightbulb lasts 1200 hours. LED lightbulbs last fifty times longer. But when they were first invented, lightbulbs could last much longer than that. In fact, at Fire Station Number 6 in Livermore, California, there's a lightbulb that has been burning continuously since 1901.

biodegradable plastics, traditional plastic production still goes up every year and is expected to grow up to 40 percent more in the next decade.

But you may have wondered why—if plastic is made from fossil fuels, and fossil fuels were once living organisms—most plastics are not biodegradable. The reason, it turns out, is similar to why nitrogen in the air is so difficult to break apart. In plastic manufacturing, carbon molecules are heated up over a catalyst, but the purpose is not to tear them apart but to lock them together to create an extremely tight bond. Once the bond is formed, the plastic becomes chemically inert. Micro-organisms that have evolved to decompose organic matter have not come across this kind of carbon bond in nature before. So, despite their billions of years of evolution, they simply do not have the metabolic pathways to digest it.[16]

While plastics can't be digested, they can be *ingested*. It is no longer news that scientists have found plastic in the guts of everything from tiny zooplankton at the base of the marine food chain to larger animals like fish, seabirds, and even whales. Every year, between five and thirteen million metric tons of plastic end up in the oceans, and every year over one hundred thousand marine mammals and over a million seabirds that swallow that plastic are killed. There is now so much plastic waste, it's been estimated that by 2050 there will be more plastic in the ocean by weight than there will be fish.

We, of course, are a part of the food chain, and plastics have been finding their way not just into our dinner plates but onto them as well. In 2015, scientists looking at fish markets off the

---

16 There are some species that are able to biodegrade plastic. The recently discovered bacterial species *Ideonella sakainesis* for instance, secretes an enzyme that under the right heat conditions can break down plastic bottles.

coasts of California and Indonesia found that one in four fish had plastic in its gut. The number was even higher in a study that looked at the English Channel, where a third of the fish caught by trawler—including cod, haddock, and mackerel—were found to contain synthetic polymers. In Scotland, a whopping 83 percent of Dublin Bay prawns, or scampi, contained plastic fibres. And in Canada, researchers found microplastics in the majority of farmed and wild clams and oysters.

But plastic is not only in the food we eat. It is also in the water we drink. In an investigation that covered over a dozen nations on five continents, 83 percent of tap water samples overall were polluted with plastic. The particles are small enough to be undetectable by the naked eye, but that does not change the fundamental truth: every day billions of people are eating and drinking plastic.

THE AIR, SOIL, AND WATER eventually become us. Ultimately, as paleoclimatologist Curt Stager writes, we are made of waste: "Look at one of your fingernails. Carbon makes up half of its mass, and roughly one in eight of those carbon atoms recently emerged from a chimney or a tailpipe. . . . [You are] built, in part, from emissions."

It's not only our bodies. Whole empires are built on waste. China's sudden rise to the status of global superpower is in large part due to its strategy of efficiently consuming North America's garbage and rebuilding itself with it. As shipping containers full of manufactured goods crossed the oceans from China and unloaded their wares on US shores, it did not make economic sense to send them back empty. Entrepreneurs took advantage of the cheap return transport to send ships back loaded with garbage and recyclable waste. America's garbage literally became China's gold . . . and silver, and copper, and aluminum, and zinc.

It is a perfect example of the old adage "One person's trash is another person's treasure." The steady supply of recycled materials was certainly cheaper than mining it at home; recycling steel, for instance, requires 60 percent less energy than mining it from iron ore. In 2010, China's primary shipments to the United States were computer and electronic equipment, valued at $50 billion. In contrast, the United States' primary exports to China, by volume, were scrap metal and paper waste—essentially, "a little more than $8 billion worth of bundled old newspaper, crushed cardboard, rusty steel and mashed beverage cans sold at rock-bottom prices," as Edward Humes writes. By 2016, China had become the world's largest net importer of garbage, taking in forty-five million metric tons of scrap metal, waste paper, and plastic from around the world each year, valued at $18 billion.

Whole towns and cities sprang up in China to recycle and profit from this waste. The town of Shijiao became the "Christmas tree light" capital, with at least nine recycling factories stripping out the copper from an estimated nine million kilograms of Christmas tree lights thrown out every year. In East China, the city of Qingdao became the primary hub for processing the world's plastic, taking in the bulk of the nine million metric tons of plastic the country imported annually. In the southeast, in the town of Guiyu, more than 5,500 businesses have been set up to dismantle over 680,000 kilograms of discarded computer equipment, cell phones, and other e-waste to harvest the precious gold, lead, and copper inside.

It was a great deal for China financially, but toxic for the country environmentally. In many towns, workers subjected to hazardous waste were found to have higher rates of birth defects, tuberculosis, respiratory problems, and blood diseases. At the same time, after four decades of being the planet's dumping ground, China had become the second largest economy in the world. As Adam Minter, author of *Junkyard Planet*, expressed in an interview on the CBC, "China's getting rich, and as you get rich, you throw away more stuff. As you throw away more stuff, you generate more for your own recycling."

China literally doesn't need to take the world's crap anymore. On January 1, 2018, it finally put a stop to toxic imports, erecting a "green fence" by banning the import of twenty-four types of *yang laji*, or foreign garbage. Almost instantly, mountains of trash that would have otherwise "disappeared" from North America began to grow in depots across the United States and Canada. The stockpiles have nowhere else to go.

The Chinese ban will force many developed nations to face their own waste or find other near-term solutions. That could mean more landfill, more incineration, or finding another country to export to. Alternatively, it could mean that we finally learn to deal with the waste appropriately ourselves.

IN THE 1985 SCIENCE FICTION COMEDY *Back to the Future*, the DeLorean time machine was truly ahead of its time: it used garbage as fuel. Today, we live in that future. Cars, bus fleets, even garbage trucks themselves, can run on biogas made from garbage. In cities like New York that generate two million metric tons of organic waste (including food waste) annually, the captured decomposing-landfill gas is being ingeniously transformed into energy.

The champions of this waste alchemy, however, are the Swedes. In Sweden, the buses run on a rather unusual mix of human waste, slaughterhouse waste, and liquor. Booze plays a part in reducing the country's emissions because it's expensive in Sweden, leading citizens to bring it back with them from journeys outside the country. Many go over the customs limit, meaning that every year hundreds of thousands of litres of beer, wine, and spirits are seized at the border. And rather than dumping the contraband down the drain, the Swedes convert it into fuel, with one litre of booze making half a litre of biogas. This "giant cocktail," as it's fondly called by the locals, powers over a thousand of Sweden's trucks and buses, along with a biogas train.

The Swedes are so good at converting their garbage that only 1 percent of its citizens' household waste ends up in landfill. Instead of burning fossil fuels for their heating, almost a million homes in the country rely on heat from incinerated garbage. As a rule, recycling stations must be found within three hundred metres of every residential area. Inside each station, high-tech vacuum systems suck up the garbage, diverting it to one of the country's thirty-two incineration plants, where it is transformed into heat or electricity. The system is so effective that Sweden now *imports* garbage, for a price, from Norway, the United Kingdom, and Ireland.[17] The country's facilities charge approximately $43 per metric ton of trash, working out to over $100 million in annual revenue.

Value can be mined from sewage as well. Every day, billions of people around the world dump tiny particles of gold down the drain, and it adds up. According to Kathleen Smith of the U.S. Geological Survey, the traces of gold, silver, and platinum that can be found in sewage sludge amount to the equivalent of a commercial mine. The precious metals, albeit in microscopic form, come from industrial waste[18] and everyday products like shampoo, detergents, and even nanoparticles that are woven into socks to reduce body odour.

The extraction method works the same way the mining industry pulls metals out of rock, by using leachates. Large-scale

---

17 As a bonus, the net CO_2 emissions from burning the waste is also negative.

18 "Precious metals such as gold could find their way into the sewers courtesy of mining, electroplating, electronics and jewelry manufacturing, or industrial and automotive catalysts."

sewage mining in a controlled environment could act to clean up waste biosolids, making sewage a better fertilizer. There is, of course, money in it too. A study at Arizona State University on precious metal recovery from sewage found that a population of one million people produces $13 million worth of metals in the wastewater annually. In Tokyo, the Suwa treatment facility has already begun extracting gold from its sewage. Rather astonishingly, the yield is even higher than it would be from mining the precious metal from ore. At Japan's Hishikari mine, which has one of the greatest gold deposits in the world, twenty to forty grams of gold on average are found for every metric ton of ore. In comparison, at the Suwa facility, 1,890 grams—almost two kilograms—of gold can be recovered from every metric ton of ash from incinerated sludge.

It would seem that our civilization is at last remembering that waste can be valuable. In Europe, engineers are using the waste heat coming from sewers and incinerators to keep buildings warm, while massive server farms like those at Google and the NSA are using cool water from flushed toilets and wastewater systems to lower the temperature at the mega-data facilities. Taboos have led to an "ick factor" that prevents most people from appreciating the vital nature of waste in the nutrient cycle. But properly treated so that it's free of heavy metals, disease, and organisms, our feces can become fertilizer or, placed in anaerobic digesters, turned into a biogas. A recent UN report suggests that human waste collected for energy has the potential value of $9.5 billion a year.

Our urine on the other hand, is already sterile. Instead of flushing it into the toxic stream of commercial, hospital, and industrial waste that pollutes fresh water, it could be diverted. The nitrogen each adult produces per year in their urine alone is enough to grow 100 to 250 kilograms of grain. Rather than relying so heavily on synthetic fertilizer, which later gets drained and dumped in the sea, we could close the loop on waste. According to the WHO, a single person, in their urine and feces combined, produces 4.5 kilograms of nitrogen every year. Scale that up with modern technology, and we can begin to return to the ancient practice followed by the Chinese, which kept the soil rich for millennia.

The mathematician Alfred North Whitehead once argued that "civilization advances by extending the number of important operations which we can perform without thinking about them." Our food comes to us from places we do not see; our energy is produced in ways we don't understand; and our waste disappears without us having to give it a thought. When we have such giant blind spots concerning our food, energy, and waste, it is no indication that we have "advanced" as a society.

Humans are no longer in touch with the basics of their own system of survival. Instead, we have systematically broken nature's grand cycles of life, death, and rebirth. By hijacking the cycle of birth and producing animals through factory farming, domesticated livestock now outnumber wild mammals 15:1. This is our food system.

By exhuming prehistoric graveyards, we have disrupted the cycle of death. Instead of leaving fossil fuels buried, we have unleashed 45 percent more carbon dioxide than would naturally be in the atmosphere. This is our energy system.

And by no longer using our waste to regrow food in our soils, we have cheated the cycle of rebirth and turned to machines to artificially suck the element of nitrogen out of the atmosphere. This is our waste system.

In turn, the human population has boomed—demanding more food and more energy and producing more waste—and soon we will be upwards of ten billion. When you hear people say that the "system" is broken, this is not usually the system they are referring to. But this is the system that sustains us on

Earth: it is our life support system. And the ruinous feedbacks caused by it are growing rapidly: overpopulation, climate change, and dead zones. Each on its own is deadly. Together, they are catastrophic.

One might say, We are ultimately in control of this system, are we not? This is *our* system, human-made. But if we can see that things are going wrong, given the time limits on our survival, why are we not dramatically changing course? In the next section, I will argue it is because there is another system keeping us in our place, one that insists on maintaining order and the status quo. In order to see it, however, we have to open our eyes to the invisible dimensions we inhabit. We need to look deep into the blind spots of time and space.

PART THREE

CIVILIZATIONAL BLIND SPOTS

WHAT CONTROLS US

7
TIME LORDS

They deem me mad because I will not sell my days for gold;
and I deem them mad because they think my days have a price.

—KAHLIL GIBRAN

ON JANUARY 1, 2018, at 12:05 A.M., several hundred passengers in Auckland, New Zealand, stepped on board a metal time machine and travelled back to 2017. It made big news around the world, though the time machine was not a new invention. It was Hawaiian Airlines Flight 446, performing its regular scheduled duty. The only difference was that on this occasion it departed shortly after midnight. It flew northeast over the international dateline back to Hawaii, which is twenty-three hours behind New Zealand, allowing the plane to land in the previous year at 10:15 A.M.

The imaginary line that separates the world into two days is known as the international date line. First drawn in 1884, the north-south line runs halfway around the world from the heart of our global time system: the prime meridian in Greenwich, England. It doesn't follow the same straight path as a line of longitude; instead, as the date line holds no legal international status, it zig-zags around countries, which are free to decide on which side they wish to belong.

Because of this, certain countries are in fact two days apart instead of just one. How is this possible? Over a hundred years ago, Samoa decided to stay "a day behind" so that it could be in the same time zone as the United States to facilitate trade. Kiribati, which is located slightly farther to the east and an hour behind, chose to remain on the opposite side of the date line, a day in the future. Then in 2011, Western Samoa exercised its sovereign right to change its mind. As Australia and New Zealand had become more important trading partners, it hopped back into the future, fast forwarding from December 29 to December 31 (skipping December 30 entirely) to join the nations on the other side of the date line. American Samoa, a territory of the United States, elected to stay behind, however. This means that for two hours each day there are *three* calendar days on Earth. If it's Tuesday at 11:30 P.M. in American Samoa, it's 6:30 A.M. on a Wednesday in Toronto, and 12:30 A.M. on Thursday in Kiribati.[1]

---

1 Time zones are also not static. In the winter for example, the time difference between Toronto, Canada, and São Paulo, Brazil, is three hours. In March—with changes to daylight savings time in the northern hemisphere (spring forward) and southern hemisphere (fall back)— it shrinks to a one-hour time difference.

Politics dictates time, but so does geography. At the North and South Poles, for instance, there are no time zones, because here all the globe's lines of longitude converge. At latitude 90° north, where the ice is constantly shifting, there are also no permanent residents, and so this Arctic spot is technically time free. Polar explorers have a few options when deciding what time to call it: they can choose to use a time that's convenient; they can use the time of their home country; or they can use Greenwich Mean Time, just like astronauts who circle the planet sixteen times a day.

The fact that everyone in the world has to set their clocks by a standard set by the Royal Observatory in Greenwich, England, is a clue that the way we inhabit the global realm of time is not something natural. It is a technology, and like many technologies it arose from a practical need.

The reason the prime meridian runs through Greenwich and not somewhere else is that this particular town was the scene of an epic eighteenth-century battle between the two timekeepers of the age: astronomers and clockmakers, who were fighting over who could claim to provide the most accurate method for timekeeping. The stargazers mapped out time from the heavens, as they had done for ages, while the clockmakers put their faith in their own hands and their ability to build timekeeping machines.

For mariners, the ability to tell the time was not a trivial issue. It was a matter of life or death. It was also a matter of national interest. In 1714 Parliament offered a prize of £20,000—millions of dollars in today's money—to the first person who could accurately chart longitude. While latitude, north or south, could be determined by the position of the sun, longitude was far more difficult to establish once land was out of sight. The night sky could be used to navigate, as it had been for centuries, but it wasn't particularly accurate, and of course navigators might want to know where they were during the day. During daylight hours, having an accurate record of the time was critical for measuring the distance a ship had travelled east or west. It was easy to see what time it was locally, but only by comparing local time with the hour back home could they know how far they were from home. And to do that, they needed a clock.

As Dava Sobel notes in her book *Longitude*, "One degree of longitude equals four minutes of time the world over, but in terms of distance, one degree shrinks from sixty-eight miles at the equator to virtually nothing at the poles. . . . For lack of a practical method of determining longitude, every great captain in the Age of Exploration became lost at sea despite the best available charts and compasses."

In the end, the problem was solved by a master clockmaker named John Harrison, who made a clock so precise that it lost only a third of a second per day. And as sea clocks, or chronometers, spread, so too did the British Empire. Thus, it was not only "guns, germs and steel," to reference Jared Diamond's book, that aided Britannia in conquering new lands, it was also the empire's mastery over time. According to horologists, the technology of chronometers empowered the British to "rule the waves" and conquer new lands beyond them.

Time, however, is not just imaginary lines. As a dimension, time may be invisible to us, but we can feel its presence. We can see the effect of time on our bodies as we age and witness its cycles on the planetary body in the seasonal flush of greens, reds, and whites. People have long relied on nature to tell the time in this way, including in the beautiful form of flowers. In 1750, the Swedish botanist and famed taxonomist Carl Linnaeus came up with a clever idea for timekeeping. He drew the plans for what he called a *horologium florae*, or flower clock. Knowing that certain plants bloom at certain times of day, he surmised that you could tell the time just by looking at a garden and seeing which flowers were in bloom.

Linnaeus called these special plants *aequinoctales*. The day lily, hawkweed, garden lettuce, and marigold were some of the species he had observed opening at specific hours. And so, a working flower clock, as poet Tom Clark imagines it, would contain a garden that looks something like this:

6 AM Spotted Cat's-ear opens
7 AM African Marigold opens
8 AM Mouse-ear Hawkweed opens
9 AM Prickly Sow-thistle closes
10 AM Nippleworth closes
11 AM Star of Bethlehem opens

Linnaeus's flower clock was never really going to catch on, because for most of the plants he observed, the time at which their flowers opened depended not on a particular hour per se but on the amount of daylight they received. Flowers are local clocks. Under the long summer sun, flowers at the northern latitude of Uppsala, Sweden, would not open at the same time as they would in Brooklyn, New York.

Sight, however, is not the only way to tell the time of day; we can also do it through sound. While we are all familiar with alarm buzzers and school bells, nature's wake-up call still comes to us in the form of birdsong. *The Horological Journal* reports that you can tell time through this "ornithological clock" if you know when specific bird species sing. For example, the green chaffinch ("the earliest riser among all the feathered tribes") sings from 1:30 to 2 A.M., the black cap follows from 2 to 3:30, next is the hedge sparrow from 3 to 3:30, the blackbird from 3:30 to 4, the larks from 4 to 4:30, the black-headed titmouse chirps in from 4:30 to 5:00, and finally the sparrow sings in the dawn from 5 to 5:30. But again, nature's clock can't be replicated, because it's both location- and bird-specific.

Beyond sound and sight, there's another sense we can use to tell time as well. During the Song Dynasty (960 to 1279), the Chinese built incense clocks to smell the hours passing. As Robert Levine writes in *A Geography of Time*, "This wooden device consisted of a series of connected small same-sized boxes. Each box held a different fragrance of incense. By knowing the time it took for a box to burn its supply, and the order in which the scents burned, observers could recognize the time of day by the smell in the air."

These are just some of the methods our species has used to tell time, but humans aren't the only creatures who are timekeepers. Many animals—like bees, rats, and cicadas, to name a few—are known to track the passage of time accurately. But one key question often arises: Are these animals telling the time based on external environmental cues like the sun, or is there some other biological clock that lets them track the time internally?

One of the best-known studies to examine how animals experience time was done by scientists Max Renner and Karl von Frisch in 1955. It was known that common honeybees often fly off to feed at the same time each day and that they could also be trained to scout for food at specific times. The researchers wanted to find out if a change in time zone would affect the bees' behaviour. So, they placed forty honeybees in a sealed room in Paris and trained them to arrive for dinner each night between 8:15 and 10:15 P.M. The researchers kept the light, temperature, and humidity in the room constant. Then one night in between feedings, Renner packed the bees up in a box and took them across the Atlantic. When he opened the box again, the bees were in an identical sealed room but this time in New York City.

The question was, when would the bees come out to feed? Was there some external cue related to Earth's position and unrelated to sunlight that would prompt the bees to feed at 8:15 P.M. New York time? The bees answered the question by emerging in the middle of the afternoon, at 3:15 P.M., proving that they were indeed tracking time internally, because it was exactly 8:15 P.M., or dinnertime, in Paris.

In the South Pacific, another creature is known for its precise timing: the palolo worm. Each year, the sea worms take part in a massive spawning event that brings millions of them to the water's surface, as they have synchronized their mating to the phases of the moon. For the locals, this orgy is also a gastronomical feast. As an article in *National Geographic* notes, "The worms are fried in oil or baked into a loaf with coconut milk and onions. A new daily special shows up on local restaurant menus: palolo worm on toast. It's considered quite a delicacy." In fact, in Vanuatu, the event is of such great importance that it is marked in the lunar calendar.

So how does an animal as seemingly simple as a worm know how to tell the time? Studying a different species of marine worm, *Platynereis dumerilii*, neurobiologist Kristin Tessmar-Raible has found evidence of a biological lunar clock. In a lab setting, using LEDs and standard light bulbs, she found that worms raised in an aquarium with the lights constantly on or off never developed reproductive cycles. But if the lights were turned on for a set period to act as an artificial moon, the worms would synch up with their own circadian clock.[2] The exact mechanism behind the behaviour still remains a mystery, but the worms do have light-sensitive neurons in their brains. Researchers believe this triggers some kind of repeating neural circuit so that "something in the body preserves the memory of those nocturnal illuminations."

---

2 That is, the worms have two clocks—one circadian (based on the day), and one circalunar (based on the month).

Of course, the human animal also has a circadian rhythm. Like most creatures, our daily cycle is synched to the sun, and while it's not a perfect twenty-four-hour match, it is still remarkably close. Harvard University researchers found that the average person's internal clock runs on a cycle of twenty-four hours and eleven minutes, with most subjects off by just plus or minus sixteen minutes.

Scientists were curious to see if the human circadian rhythm could be tricked in the same way as the bees. To find out, in 1972, a geologist named Michel Siffre agreed to take part in a NASA-funded study where he would spend six months alone in a cave in Del Rio, Texas. The goal? To discover how the human body would respond to long-term isolation. Siffre would not starve in the cave. He had his basic needs taken care of: plenty of food and water, and he could even control the temperature and artificial lights. Aside from that, though, he had no external cues like sunlight or the seasons to mark his time inside.

As a seasoned cave explorer, Siffre found the first two months relatively easy. He read Plato, listened to records, and explored his new surroundings. As a part of the study, electrodes were attached to his body to monitor his brain, heart, and muscle activity. He also kept a journal detailing his experience. One of the key findings was that without daylight to calibrate his internal clock, Siffre's body loosened itself from the twenty-four-hour day and found a different cycle of time. Periodically, he would stay up for thirty-two hours and sleep for sixteen. And a couple of times, his body tuned to a forty-eight-hour cycle for a while, though the range generally shifted from eighteen to fifty-two hours.

Inside the cave and without much light, time's boundaries began to erase. "I believe that when you are surrounded by night—the cave was completely dark, with just a light bulb—your memory does not capture the time," he said. "You forget. After one or two days, you don't remember what you have done a day or two before. The only things that change are when you wake up and when you go to bed. Besides that, it's entirely black. It's like one long day."

By day seventy-nine, Siffre had not only begun to lose track of time, he had begun to lose his mind. With no outside contact, the desolation he felt was overwhelming. His only "friend" was a mouse that came to steal his supplies. So it came as a terrible loss when one day, he accidentally crushed the mouse (he was attempting to catch it with a casserole dish to make it his pet); now, with nobody around to keep him company, his depression worsened. Siffre began to contemplate suicide. He became so steeped in this mental darkness that when a lightning storm outside sent electricity coursing through his cardiac electrodes, he was so psychologically numb he let it happen four times before he even thought to remove the wires.

Finally, on day 179, he was brought out. The study had been torturous, but if there was one small blessing it was this: Siffre had thought that less time had passed in the cave. In subsequent studies it's been found that isolation from external cues dilates time. Siffre thought he had been in the cave for only 151 days.

AS A DIMENSION, time reaches far beyond human scale and perception. That's because the true nature of time is not only deep but infinite, stretching from the Big Bang right down to the present moment. In Hindu and Buddhist cosmology, the Sanskrit word *kalpa* refers to this elongated time, with each *kalpa* lasting an eon of 4.32 billion human years. This stretching of time shifts our perceptions[3] and in a way gives us a better

---

3 The Long Now Foundation is currently building a ten-thousand year clock in Nevada that will only tick once a year. Its purpose is to encourage people to consider a long view when it comes to the nature of time.

sense of where we lie on the continuum. Imagine, for instance, if instead of the year 2020 we marked the date from a different point of origin. That is, if we counted not from the birth of Christ but from the birth of the solar system. How differently would we see our time on Earth if we wrote the date as January 26, 4.543 billion?

On the hyper-immediate scale, we also speak of quick time. If you hear someone say "I'll be there in two shakes" or "in a jiffy," colloquially that means they will be there soon or right away. But a "jiffy" in electronics has a more precise meaning. It refers to the period of an alternating current power cycle, or 1/60 of a second. A "shake," on the other hand, is defined as ten nanoseconds, or 10^{-8} seconds, and is used in physics as a measure of timing chain reactions in a nuclear explosion.

For non-scientists, what we generally speak of when we refer to time is what we experience on a human scale. For most of history, time was somewhat tangible. Our own bodies kept track of day and night, and stellar bodies marked our seasons and astronomical calendars. The sun, the biggest and brightest star in our sky, is the solar body we still use to mark the passing of time. For the ancient Egyptians, when the sun went down, measurement of time itself disappeared, and this remained the case with Roman sundials. As the motto on one sundial read, "*Absque sole, absque usu*," or "Without sun, without use."

The water clock, or clepsydra, was the first device that marked the passage of time after sunset, using a measured quantity of water that took a known period of time to drip through a hole in a vessel. Hourglasses, used primarily for decoration today, similarly keep time using gravity.

In different parts of the world, time has been measured in ways that we might now consider rather peculiar. As historian E.P. Thompson notes, "In Madagascar time might be measured by 'a rice-cooking' (about half an hour) or 'the frying of a locust' (a moment). The Cross River natives were reported as saying 'the man died in less than the time in which maize is not yet completely roasted' (less than fifteen minutes)." Time was not an abstract construct but a real measure framed by the typical duration of events. But all timekeeping methods of the past—from using the stars to water, and from cooked rice to fried locusts—were still measures of the transformation of physical bodies. It was not until the invention of the clock that time became a measure entirely unto itself.

Today, if you look down at your wristwatch, you are seeing a time that the watch itself created. That is, clock time is a human invention. That's why we still say "of the clock," abbreviated to "o'clock." The distinction between clock time and lived time used to be important, but today "o'clock" time is almost universal. Few of us measure the time of day by the rise and fall of tidal rivers or the placement of the stars. Time is no longer a dimensional flow that we, along with the rest of nature, inhabit but rather a construct, a "thing," that orders our lives and that we must obey.

That is not to say that having coordinated time is a bad idea. Before we were in synch, appointments were hard to schedule and most meetings had to be made at a clear and specific time, like dawn. Time being inexact in the past also meant it was flexible. And in many parts of the world today, you will find that time is still not a rigid entity. That's one of the reasons why people love the experience of "island time." For travellers, it takes over from the rigorous pace of the modern world. Here, time does not control the locals, because it is the locals who control time.

Localized time is how we experienced time for most of history. The first mechanical clocks appeared around the early fourteenth century and used something called a verge escapement. In essence, a weighted mechanism shifts back and forward to move a toothed wheel. This created the tick, or the beat, of time.

These early clocks were placed in town centres and churches and signalled the timing of public events. A century later, in the early 1500s, German locksmith Peter Henlein invented the first watch. He also became the manufacturer of the first pocket watches, in 1524. These portable clocks shrank time devices down in a manner similar to how personal computers shrank to the size of cell phones. And like the first cell phones, these new clocks were expensive. Henlein's creations were affordable only for the rich.

It was not until the early 1800s that portable timepieces entered the mainstream. To determine longitude, mariners became the first early adopters of portable chronometers. In 1737, only one chronometer existed, but by 1815 there were over five thousand. It was in fact the military that popularized the use of wristwatches. In 1880, Swiss watchmaker Constant Girard mass-produced two thousand wristwatches for German naval officers. By the First World War, these wristwatches, or "trench watches," as they came to be known, allowed soldiers to coordinate their movements without having to rummage through their rucksacks to find a pocket watch. For aviators, a wristwatch was literally handy, as it allowed them to keep both hands on the controls.

Time, though, still kept its own local pace around the world. And the pressure to create a global time based on European measures met with resistance. As Ian Beacock writes in *The Atlantic*, in his article "A Brief History of (Modern) Time," changing local time to a standardized time in some cases led to violent opposition:

> In January 1906, several thousand cotton-mill workers rioted on the outskirts of Bombay. Refusing to work at their looms, they pelted factories with rocks, their revolt soon spreading to the heart of the city, where more than 15,000 citizens signed petitions and marched angrily in the streets. They were protesting

the proposed abolition of local time in favour of Indian Standard Time, to be set five-and-a-half hours ahead of Greenwich. To early 20th-century Indians, this looked like yet another attempt to crush local tradition and cement Britannia's rule. It wasn't until 1950, three years after Indian independence, that a single time zone was adopted nationwide. Journalists called this dispute the "Battle of the Clocks." It lasted nearly half a century.

Time, as we shall soon see, has come to benefit some people at the expense of others. And as timekeeping has become more precise, modern time has become less flexible. Having lost the connection to measures we can physically perceive, modern time is not a measure of bodies or events. It is imperceptible. And this is our blind spot: we have confused our human measurement of time with time itself. Time's power over us is that it is intangible. After all, how can you control something you have no control over?

The average quartz watch oscillates based on the imperceptible vibrations of a crystal at 32,768 Hz. Atomic clocks that govern everything from GPS to our smartphones to our traffic lights measure the invisible vibrations of strontium atoms. The United States Naval Observatory's atomic clocks are so accurate that they won't lose a second over three hundred million years, but even this level of accuracy has nothing on the latest horological devices. Today, the world's most precise atomic clock will lose only one second every ninety billion years. Time as a measurement itself is now utterly abstract. As Dean Buonomano writes in *Your Brain Is a Time Machine*, "In 1967 an international consortium defined a second as: 'the duration of 9,192,631,770 periods of the radiation corresponding to the transition between the two hyperfine levels of the ground state of the caesium 133 atom.' The basic unit of time became permanently divorced from the observable dynamics of the planets and placed in the domain of the imperceptible behavior of a single element."

If time was once subjective, today it is completely objective. It is not the sun that tells us the time of day. Instead, we march to the drumbeat of atomic time. All of humanity has been synchronized. Time is no longer a signal from the sun but from a constant pulse beamed at us from satellites in the sky.

IN 2013, MIWA SADO, a thirty-one-year-old reporter for the broadcaster NHK in Japan, died suddenly, cell phone still in hand, from a condition known as *karoshi*, a Japanese term that means "overwork death." The term exists because in Japan occupational mortality is a real condition. Sado had put in 159 hours of overtime[4] that month before collapsing due to congestive heart failure. According to Tokyo-based reporter Jake Adelstein, her sudden passing was one of thousands of deaths from suspected overwork that take place in Japan every year.

The government launched an investigation and in 2016 issued its first white paper on *karoshi*. The report found a full one-fifth of Japan's workers were at risk of overwork death. According to the companies surveyed, 22.7 percent of their employees logged over eighty hours of overtime per month—the level at which overtime becomes a health risk—and a further 12 percent logged over one hundred hours of overtime.

4 This is not a situation unique to Japan. Charles Czeisler, a professor of sleep medicine at Harvard Medical School has documented sleep deprivation in hospital interns who are at times scheduled to work twenty-four to thirty-four hour shifts. A lack of sleep presents real dangers: sleep-deprived interns made 36 percent more serious medical errors and 5.6 times as many diagnostic errors. They also increased their risk of stabbing themselves with a scalpel or needle by 61 percent.

The tyranny of time is not new, and certainly not unique to Japan. Time has been structuring our days for millennia. Even two thousand years ago, people complained. Writing in 224 BC, the Roman playwright Plautus famously cursed the sundial in one of his plays:

> The gods confound the man who first found out
> How to distinguish hours. Confound him too,
> Who in this place set up a sundial,
> To cut and hack my days so wretchedly
> Into small pieces! When I was a boy,
> My belly was my sundial—one surer,
> Truer, and more exact than any of them.
> The dial told me when 'twas proper time
> To go to dinner, when I ought to eat:
> But nowadays, why even when I have,
> I can't fall to unless the sun gives leave.
> The town's so full of these confounded dials.

Yesterday's sundials are today's digital clocks. And while we are often reminded that our ancestral past was brutal, it is worth remembering that in sourcing food the average hunter-gatherer adult only worked three to five hours each day, or twenty hours a week. Materially, we may have been poor, but temporally we were rich. And if time really is money, we may not have been so badly off.

The sociologist Daniel Bell made the keen observation: that "industrialization did not arise with the introduction of factories, it "arose out of the measurement of work." Said another way, it wasn't the invention of the steam engine or the spinning jenny that changed the world so much as our attitudes towards time. In a sobering case of "be careful what you wish for," it turns out that the idea of overtime came from European cloth makers in the 1300s. In the Middle Ages, labour time was, as it had always been, equivalent to the amount of work that could be done in the light of day, from sunrise to sunset. Still, it was rather leisurely.[5] In towns, church bells signalled the beginning and end of the work day; for agricultural labourers, there was no need for such a prompt, because a field could not be worked in the dark. Cloth makers, on the other hand, worked indoors in the mills, and because there were many festive days in a year (marked by the religious calendar), they were forbidden from weaving during those times. That meant that they couldn't always get the job done. In a plea to complete their work, they began to demand longer hours and higher wages, something all too familiar to us today but unknown at the time. Working at night, though, was illegal; those caught working by candlelight were fined and banished from the trade for life.

It was only in 1315 that mill owners began to authorize night work. This was, as far as we know, one of the first instances of being paid by clock time. In the cloth towns of Artois and Flanders, as historian Peter Stabel notes, we began to see the emergence of the term *clocke des ouvriers*, or "bell of the workers." And soon a new rhythm would be set by textile manufacturers.

Unlike the previous church clocks, the workers' new adherence to clocks, or *Werkglocken*, was more strictly enforced. To prevent them from cheating and using their time idly, the new clocks signalled when weavers should go to work, when they should break for lunch, when they should finish eating and return to work, and when they should quit at the end of the day. The weavers didn't like the new arrangement nearly as much as they thought they would. Fines were issued to weavers who arrived late to work after the morning bell, and efforts to resist

---

5 In fourteenth-century England, peasants worked around 150 days per year.

the new bell-regulated time were penalized. In the most extreme instances, using the bell to call together an armed workers' resistance or a revolt against the king, alderman, or bellsman was punishable by death. As historian Jacques Le Goff notes, this temporal transformation marked a threshold between the world of ancient, natural cycles and the world we know today: "In the cloth manufacturing cities, the town was burdened with a new time, the time of the cloth makers."

In the Middle Ages, it was difficult to incentivize weavers to work themselves to death, because hoarding pay beyond what was necessary to live wasn't particularly worthwhile. It wasn't until the industrial era brought in a fledgling consumer society that additional pay offered the promise of greater luxuries and the possibility of upward mobility. But time has always been regarded as precious, and people would not give it up easily.[6] Workers had to be trained.

The late seventeenth century is the first time the word "punctual" enters our vocabularies in the way it's used today. Before then, the word had a different connotation and implied that a person was a stickler for details. Being "on time," however, was increasingly touted as a virtue. In a pamphlet called *Friendly Advice to the Poor*, the English Reverend J. Clayton complained of "idle, ragged children" who did not obey the time. He advocated for a new purpose in schools. In 1755, he wrote, "the Scholars here are obliged to rise betimes and to observe Hours with great Punctuality." School bells were introduced to structure the day, because a factory model of repetitive order prepared the young for hard work and industry. As Alvin Toffler writes in *Future*

---

6 A survey of 1,018 full-time U.S. employees found that 41 percent of people prefer having time to money. That said, only 30.3 percent of people surveyed were willing to give up their present salaries for a better schedule.

Shock, "Children marched from place to place and sat in assigned stations. Bells rang to announce changes of time. The inner life of the school thus became an anticipatory mirror, a perfect introduction to industrial society. The most criticized features of education today—the regimentation, lack of individualization, the rigid systems of seating, grouping, grading and marking, the authoritarian role of the teacher—are precisely those that made mass public education so effective an instrument of adaptation for its place and time."

This was a culture making a transition to a machine-based manufacturing world. It was an epic shift. And shaming those who abused time was one way to implement it. By the eighteenth century, punctuality and exactitude were encouraged as hallmarks of good citizenry, while slothfulness and "time-thrift" at work were seen as traits of the poor and slovenly.

It was around this time that Benjamin Franklin famously declared that time is money. In a 1748 essay entitled "Advice to a Young Tradesman," he wrote, "Remember that time is money. He that can earn ten shillings a day by his labour, and goes abroad, or sits idle one half of that day, though he spends but sixpence during his diversion or idleness, it ought not to be reckoned the only expence; he hath really spent or thrown away five shillings besides."

Industrial capitalists now "owned" working time, and just like that they owned the workers. Once you've sold your time, you can hardly spend it as you like. In the eyes of the managers, the unprofitable use of time was theft, so the habits of wasting time had to be bred out of the workforce. One way to do that was with penalties. In England's County Durham, Sir Ambrose Crowley and his son wrote out ninety-four rules in *The Law Book of the Crowley Ironworks*, which explicitly dealt with time. Wasted time was unpaid time: "This service must be calculated after all deductions for being at taverns, alehouses, coffee houses, breakfast, dinner, playing, sleeping, smoaking, singing, reading of news history, quarelling, contention, disputes or anything foreign to my business, any way loytering."

Factory time, however, was not to stay within the confines of the factory or the school. It was spreading. Having gone from being a vast dimension to a "value," time was soon to become a "product" in itself.

IN THE 1800S, time was haphazard. In 1875, American railways had to operate under seventy-five different local times across the country, with most towns still structuring their local time by high noon, when the sun was at the highest point in the sky. In Germany, as Ian Beacock writes, "travellers had to clarify whether departures were according to Berlin, Munich, Stuttgart, Karlsruhe, Ludwigshafen, or Frankfurt time." To avoid all this confusion, as is well known by now, the synchronization of time and the creation of time zones was initiated in large part by the railways.

The problem wasn't the clocks; by now they were good at keeping local time. But without standardization on the national or continental level, they were effectively useless when it came to coordination. And so two entrepreneurs quietly stepped into this temporal fray and effectively changed forever how we perceive time.

Samuel P. Langley was the first person to sell time. In 1867, as the director of the Allegheny Observatory in Pittsburgh, Langley convinced local businesses and industries to pay for his time signals. His "master clock" would transmit the time via telegraph so that clients' "slave clocks" could synch up to match it. Companies like Western Union and the Pennsylvania Railroad were soon on board, with the latter paying $1,000 a year for the observatory's time signals.

Leonard Waldo, an astronomer and director at Yale University's Winchester Observatory, took it a step further. Believing scientists could do a better job, he promised to sell a more exact time than was offered by Langley (which was off by one to two seconds). What both men had in common was that they became proselytizers for this new time and began calling for the end of local time, which Langley declared a "fiction" and a relic of the past.

This new time, however, would require a re-education of the masses. Writing to the railroad commissioners, Waldo stated that "any service which will train these persons into habits of accuracy and punctuality, which will affect all employers and all employees with the same strict impartiality, so far as wages for time deployed is concerned, will be a great benefit to the State."

In 1891, the Electric Signal Clock Company began selling a factory clock aptly named the Autocrat. The new system, the brochure proclaimed, "gives military precision, and teaches practicality, promptness and precision wherever adopted . . . providing management and supervisors *a means to extend their disciplinary reach beyond their vision* [emphasis mine]."

In 1893, Bell and master/slave clock systems were soon installed in factories. At the Chicago World's Fair, a new factory clock was unveiled: one master clock connected to two hundred slave clocks, which were programmed to ring bells so that workers would start and stop machines.

For management, time control became a new form of power. The watchmakers, eager to sell their wares, also got into the action, promoting punctuality as a virtue and tardiness as undisciplined and undesirable. As Robert Levine writes, watchmakers started marketing the idea that it was important to "keep a watch on everybody" both literally and figuratively, an augury, we shall soon see, of what was to come. Punch clocks were also invented around this time to mark the exact moment each worker came to and left work. Within a few decades, all the time-stamping companies came under the consolidation of William Bundy, who created the International Time Recording Company, later known as International Business Machines, or IBM.

Today, this "technological conditioning" has reached a point where business time has so infiltrated our minds that we no longer need punch cards to know the importance of arriving on time at our jobs. Billions of people around the globe now wake up each morning to alarms, commute en masse to work, and arrive and depart at the scheduled times without thinking about why or how they came to be trained for the task. From the invention of punctuality and the new insistence never to be "behind the time," factory time broke us away from nature's cycle. It may have taken several generations, but as E.P. Thompson writes, "In all these ways—by the division of labour; the supervision of labour; fines; bells and clocks; money incentives; partings and schoolings . . . new labour habits were formed, and a new time-discipline was imposed." We now inherit this idea of institutionalized time and teach it to each new generation. Although it was completely (and recently) manufactured, it allowed machine time to flow into the present century.

TODAY, MACHINE TIME is everywhere. Time is still money, but in places like India or Southeast Asia—where sweatshop labourers have supplanted European weavers—it's not much money. If you can buy a nice shirt online for $13, then it isn't only because of technology. It's because in Bangladesh there is someone sewing it together for just US$.013 an hour.[7]

---

7 Garment workers in Bangladesh are some of the lowest-paid workers in the world. On average, a monthly salary is $68, which is significantly below a living wage. Workers often work seven-day weeks, with overtime amounting to fourteen to sixteen hour work days.

That same shirt in the Middle Ages, historian Eve Fisher argues, would have cost thousands, if medieval workers were paid a modern American minimum wage. She calculated that given the time it took to spin the fabric (480 hours), weave it (20 hours), and then sew the material together by hand (8 hours), the average shirt required 508 hours of labour. In the United States, where the federal minimum wage is $7.25 an hour, producing that shirt by hand would cost $3,683. In Bangladesh, the shirt could be made for $65.

Of course, even with cheap labour, we don't do all of our work manually. In fact, it was precisely because making cloth was so time-intensive that with the Industrial Revolution spinning machines were among the first inventions, and textiles were the first products to be mass-produced. The spinning and weaving of fabrics is now all done by machine, and at the end of the line is the factory worker, also working with a machine, who puts in the final hours of sewing. It is all this *invisible labour*, then, all of this *invisible time*, that transforms a shirt that once took 508 hours to put together and makes it available for a fraction of what the person who wears it might earn in a day.[8]

Working faster, and shaving off more time equals more profit: a practice that has consistently gained momentum since Frederick Taylor introduced "scientific management" in the early twentieth century; the idea of breaking down tasks into individual motions. By the 1960s management experts optimized efficiency even more by breaking office and factory work right down to the minute. A manual published by the Systems and Procedures Association of America offered this "universal standard data" to employers, giving them a sense of how long basic tasks took:

---

8 This author found a "poet style" similar top with yoke on sold by H&M for $13 that was similar to the type of shirt described by Fisher.

> Open and close file drawer, no selection = .04 minutes; desk, open centre drawer = .026 minutes; close centre drawer = .027 minutes; close side drawer = .015 minutes; get up from chair = .033 minutes; sit down in chair = .033 minutes; turn in swivel chair = .009 minutes; move in chair to adjoining desk or file (4ft max.) = .050 minutes.

By the 1980s, as Andrew Goatly notes in his book *Washing the Brain*, this dehumanizing tempo had entered the sweatshops, where "25% to 30% of all clerical workers found themselves being supervised by computer in electronic sweatshops doing boring, repetitious, fast-paced work that requires constant alertness and attention to detail."

Fast forward to today, where workers in factory and office settings are routinely surveilled and timed for productivity. At Pegatron, China's second largest manufacturer for Apple, "The workday typically lasts 12 hours on the assembly line. There are 90 minutes of breaks for meals and restroom. No talking. No standing up. No drinking water at your station. No cell phones. If you finish your work early, you must sit down and read employee manuals. . . . The day's task is to assemble back covers for the iPad. The quota is 600 per day, or 1 per minute."

And it's not just China. At Amazon warehouses in the United Kingdom, an undercover investigation revealed that 74 percent of workers were afraid to use the toilet because it would affect their target numbers. Many chose instead to urinate in bottles. At Centrelink, a call centre in Australia, employees punch in an ID code each time they want to use the bathroom. Breaks are timed, and workers are reprimanded for anything over five minutes. Bathroom breaks may seem trivial, but they are a pretty clear signifier of the way we all lose dignity when our time is not our own. According to a three-year investigative report by the charity Oxfam, titled *No Relief: Denial of Bathroom*

Breaks in the Poultry Industry, workers in many US processing plants live under such a "climate of fear" that they dare not even *ask* to take a break. Instead, "Workers urinate and defecate while standing on the line; they wear diapers to work; they restrict intake of liquids and fluids to dangerous degrees; they endure pain and discomfort . . . [that put them] in danger of serious health problems."

Losing minutes means losing money. But in our modern system, a minute is a lifetime. Even microseconds count. And nowhere has the pace set by machines outstripped human ability to keep up more than in the global financial markets. Wall Street trades on a time scale that the human brain cannot even perceive. In the past, merchants trekked for weeks and months in order to trade their wares, whereas today's trade in stocks is a shuffle of billions of "buys" and "sells" sealed by computer handshakes that happen at the speed of light.

These days, we routinely send round-trip signals from New York to London and back again in about sixty milliseconds. That means I could send a financial transaction across the Atlantic Ocean six times in the time it takes you to read this sentence. If time has always been invisible to us, time in the past at least registered at a human scale. We could see the shadows of the sundial or watch the minute hand tick from one second to the next. For high-frequency traders working in today's stock markets,[9] the financial time they trade within is completely undetectable. The markets operate in what Jeremy Rifkin calls "computime," a blur of digital time that flickers past us at such high speeds that the clocks can no longer be read by humans; they can only be understood by computers.

---

9 High-frequency trades now account for 50 to 70 percent of stock market volume.

For us, one millisecond[10] passing to the next is imperceptible. In fact, it takes the human brain a whole thirteen milliseconds just to process an image. But for computers, it is these briefest of gaps of time within the network where algorithms make split-second financial decisions that are exploited for a profit. As Sal Arnica, author of *Broken Markets*, has said, "By the time the ordinary investor sees a quote, it's like looking at a star that burned out 50,000 years ago." That said, financial time does not operate evenly in the world. There is computer time and human time. We have just looked at how clock time is utilized to trade within the global marketplace, but our worth—and how we much we sell our time for—more often than not has less to do with our intelligence, work ethic, or inherent abilities and more to do more with *where* on the planet we are born.

Time is precious to all human beings. Certainly, my time is not worth any more or any less than yours, but the rewards for our time are not equal. At the extremes of the spectrum, the difference is chasmic. In 2018, billionaire Jeff Bezos made $8.96 million an hour, even when he slept. For a Dalit—formerly known as an "untouchable" in India's traditional caste system—the dirty work of cleaning up latrines earns about forty-six rupees a day, which, given an eight-hour day, is about five cents an hour.

ONE UNEXPECTED CONSEQUENCE of being a publicly acknowledged genius is that you never seem to have enough time. Reporters were always hounding Albert Einstein to question

---

10 The human brain needs thirteen milliseconds to process an image, and it takes between one hundred and four hundred milliseconds to blink an eye. A high frequency trade travelling round-trip from Chicago to New Jersey takes only 13 milliseconds. That means 30 trades can take place in the blink of an eye.

him about his mind-bending ideas. But time is as precious for theoretical physicists as it is for CEOs, and Einstein had an ally in his long-time secretary, Helen Dukas. Whenever someone called or dropped by wanting Einstein to explain his theory of relativity, she was instructed to tell them, "An hour sitting with a nice girl on a park bench passes like a minute, but a minute sitting on a hot stove seems like an hour. That's relativity." Einstein's realization was that there could be no such thing as "correct" time at all.

Until the beginning of the twentieth century, scientists had been under the working impression that time was absolute. Sir Isaac Newton's famous 1687 treatise *Mathematical Principles of Natural Philosophy* had been the guiding principle of physics. In it, Newton outlined his three laws of motion, alongside his principles of absolute space and time, explaining how the universe functioned. We now call this Newtonian mechanics. In the Newtonian world view, if a clock were precise enough, and the same standards were set in place, then two clocks—say, one on Earth and one on Jupiter—would tick at the same pace, and thus the passage of time would be identical in both places. That seems reasonable. After all, that's how English navigators figured out longitude.

What Einstein proposed, and mathematically proved, was that this was not at all the case. Firstly, space and time are not separate things; they are unified. And because of this, any movement through space affects time. In this view, the two clocks, on Earth and Jupiter, which are moving relative to each other at different rates of speed, would tick differently and display different times on their dials.

We know this to be true, as today's GPS signals require at least four different satellites to determine a location on Earth. The atomic clocks triangulate your position by measuring the time difference it takes for the signal to arrive from the various satellites. For civilian GPS, these time signals allow us to measure the latitude, longitude, and altitude of something like your cell phone in a shopping mall down to an accuracy of 4.9 metres. But for GPS to be accurate, the clocks themselves have to be super-precise and operate within the same range of forty to fifty nanoseconds. But here on Earth, and twenty thousand kilometres up in the sky, time is not the same.

Satellites zoom around us at an orbital speed of about fourteen thousand kilometres per hour, fast enough, according to Einstein's theory of special relativity, to make their clocks run slower. Seven microseconds slower per day, in fact. At the same time, because of their distance from Earth's mass, clocks that are farther away from its gravity feel less of its effect and run faster. According to Einstein's theory of general relativity, this is the case. Without calibration, the atomic clocks housed in satellites would run *ahead* by forty-five microseconds each day. Together, then, the effects of special relativity and general relativity mean that the satellite clocks (seven microseconds behind and forty-five microseconds ahead) should be off by thirty-eight microseconds, or thirty-eight-millionths of a second each day.

Thirty-eight microseconds does not sound like much, but if we did not account for it, within minutes the error would render GPS useless. According to Richard Pogge, professor of astronomy at Ohio State University, after just two minutes spatial positions would be totally off, and with cumulative effects GPS systems would have errors in the order of ten kilometres per day.

So we know it's not just a theory. The math is accurate. And while advanced physics can account for these differences and translate them in such a way that we understand how they affect us in everyday life, physicists themselves aren't really sure if time is even a "thing" in itself; that is, if time is something we've concocted or if time as we know it really exists.[11]

Consider this: in the same way that math has "imaginary numbers" (like $i^2 = -1$, more a concept than a visible thing), physicists also work with imaginary time. Real time is what we measure in our daily lives with clocks. As Stephen Hawking said, this is "the time that we feel passing, the time in which we grow older." If real time starts with a point called the Big Bang, and ends with a point called the Big Crunch, then in imaginary time there is a time that wraps beyond those time scales—the time outside of those times—that allows us to question whether time as we know it functions according to "reality." In *A Brief History of Time*, Hawking puts it like this:

> This might suggest that the so-called imaginary time is really the real time, and that what we call real time is just a figment of our imaginations. In real time, the universe has a beginning and an end at singularities that form a boundary to space-time and at which the laws of science break down. But in imaginary time, there are no singularities or boundaries. So maybe what we call imaginary time is really more basic, and what we call real is just an idea that we invent to help us describe what we think the universe is like.

For physicists, then, imaginary time is something that "exists" in that it can be described mathematically even though it can't be experienced. The inverse could be said of clock time. We experience it even though we are not sure to what degree it really exists. That's because we invented a tool of measurement, the

---

11 Some physicists, for instance, have suggested that the "universe is timeless."

clock, that describes the measurement itself. A clock measures time in the same way that a ruler measures space. But a ruler doesn't really measure space at all; it measures intervals—or what we call centimetres or inches—measurements, as we'll see in the next chapter, that we made up.

Like the stone that Samuel Johnson kicked in Chapter Two, the realness of time, just like the realness of the stone, may be deceptive. In a similar vein, the eminent physicist Brian Greene put forward this conundrum: What if time isn't real and is just a mental projection? That is, maybe there isn't some invisible border between the past switching into the present and then switching us into the future. Maybe time isn't somewhere "out there," because time is a perception, a projection from within our brains. As he writes in *The Fabric of the Cosmos*, "This . . . leaves open a pivotal question: Is science unable to grasp a fundamental quality of time that the human mind embraces as readily as lungs take in air, or does the human mind impose on time a quality of its own making, one that is artificial and that hence does not show up in the laws of physics?"

In other words, what is science measuring if the only thing humans are even capable of experiencing is the eternal moment that we call now?[12]

There is one thing we can say for certain: whether time as a dimension or a perception is real or not, we are living by something manufactured. Einstein proved that there is no universal beat of time, and yet here on Earth we have organized our lives into a system of time that is synchronized.

---

12 Even our idea of "now" operates on delay. According to scientists, the psychological present is only three seconds long and "our consciousness lags 80 milliseconds behind actual events." According to neuroscientist David Eagleman, "When you think an event occurs it has already happened."

"IMAGINE VISITING YOURSELF in the future," the voice in the ad begins. In the scene, a frustrated man is running late and misses the bus to work. Suddenly—through TV magic—he is transported into the future, where he comes face-to-face with his older self. The two versions of himself are jogging together along a gorgeous beach on a sunny day. The older man is tanned, relaxed, and smiling. He turns to his younger self and says, "Still in the rat race?"

The ad is well known to many Canadians. In the 1990s, it was a commercial for London Life Insurance's Freedom 55, a financial planning service so named because fifty-five represented an ideal retirement age. By 2010, however, with inflation and rising costs, retiring even at the standard age of sixty-five began to look increasingly unlikely. A Sun Life Financial survey that year found that only 28 percent of Canadians believed retirement at age sixty-five was feasible. Expectations continued to change, and the following year newspaper headlines started asking, "Is Freedom 75 Boomers' New Goal?" And by 2017, a new tagline entered the mainstream that bumped up the retirement age again, with "Freedom 85: Staying in the Workforce." The thought of early retirement for many people today is a far-fetched dream. Some might wonder, "What's next, Freedom 100"? Or perhaps "Freedom When You're Dead"?

Retirement is the cheese dangled before us in the rat race. Our personal time, the thing we are all born with, is subjugated to clock time, the thing we invented. We're promised that if we work hard all our lives, in our senior years we'll be rewarded with free time to finally do exactly as we please. But as anyone who has been unemployed for a long stretch knows, leisure time is not "free" time at all. That's because leisure is sold back to us. Leisure costs money.

Even if you have all the time in the world, there are very few things that can be done on an empty pocket. You can go hiking, swimming, play a game of chess, or read a book from a library, but most leisurely pursuits have been commodified and become big industries. Yoga practitioners in the United States alone spend $16 billion on classes, clothing, equipment, and accessories. Golf, even more upscale, is a $70-billion industry in the United States. The illusion of *free* time is the ouroboros of capitalism. Leisure has entered the marketplace so that it feeds back into the economy.

Consider how shopping, an activity regarded largely as a chore until the 1950s, became a hobby, or what's now even known as "retail therapy." Today, going shopping is considered a treat, a relaxing way to spend the weekend. Beyond the Sunday stroll along the high street or at the mall, now hordes spend their days on "shopping tours," the purpose of which is not to see any sights but to be bused to windowless outlet warehouses to get deals on brand-name merchandise.

We shop as if our happiness depends on it, because we are told it does, and because we don't want to commit the sin of being behind the times. Fashion, we are told, keeps us up to date. If in the 1950s a style of skirt could last a decade, by the 1980s fashion had transformed into distinct "seasons." But for a time, these seasons were still connected to the physical seasons; we need warmer clothes in the winter and lighter outfits in the summer. On the runways, February and March are traditionally when the fall/winter collections are unveiled, and in September and October, models are dressed in the spring/summer collections for the following year. But today's fashion cycle has accelerated far beyond that. In the world of "fast fashion," there are fifty-two seasons a year. As soon as something is in, it's out, sending fashionistas scrambling to buy more clothes to stay current.

If we are trained to become slaves to fashion, we cannot forget the slaves *of* fashion: the invisible garment workers putting in overtime at fractional wages so that retailers like the global chain Zara can receive shipments twice a week and companies like H&M and Forever 21 can introduce new styles every day.

This hyper-speed manufacturing cycle also affects the natural world. In the documentary *RiverBlue*, fashion designer and activist Orsola de Castro can be overheard saying, "There is a joke in China that you can tell the 'it' colour of the season by looking at the colour of the rivers." The river behind her is stained from textile dyes. It is not blue but magenta.

The vast amount of power and energy required to mass-manufacture, market, and ship goods at this accelerated pace for a population of over seven billion has led to a very real and physical shifting of the world's weather patterns. It is no joke that capitalism's hyper-acceleration through "fashion" is literally changing the seasons. Our economy is causing climate change. For economists, this equals growth; for ecologists, it equals devastation.

Alongside the rapid rise in wildfires and deadly hurricanes, rising temperatures have severe consequences for species that migrate and feed based on the seasons. In nature, as we saw earlier, timing is everything. But in recent years, while the atomic clocks that synchronize our modern world have become ever more precise, something strange is happening to nature's clock, because everywhere around us its timing is off.

IT WAS A REGULAR WEDNESDAY AFTERNOON in Central Park, New York City. Joggers took to the paths in T-shirts and shorts, people picnicked, little kids rolled and tumbled on the grass, their faces dripping with melted ice cream. It was 25.5°C (78°F), which would have been perfectly natural for a summer's day, but if you took a moment to look around you would notice a clue that something wasn't right. Everywhere in the park, the deciduous trees were leafless. This hot summer-like day in 2018 wasn't taking place in June, it was February.[13]

Freakish weather events are occurring at an increased rate around the world. But it's only when a sustained pattern forms that scientists can attribute it to a change in climate. For gardeners everywhere, this change is increasingly evident. They only need look in their backyards to notice their own "flower clocks" have begun to bloom at unusual times. According to a major study by the Botanical Society of Britain and Ireland, plants are blooming ahead of their spring schedule. In 2016, over six hundred plant species bloomed early, almost double the number of the previous year. Across the Atlantic, the same trend has been documented. In 2010 and 2012, plants on the east coast of the United States were flowering earlier than at any point in recorded history. According to scientists, it's happening because the "sweet spot" temperature at which seeds begin to develop has been coming earlier. For plants like *Arabidopsis thaliana*, that temperature is between 14°C and 15°C. Early warm spells are affecting many spring flowering plants, with every 1°C rise in temperature triggering plants to bloom on average 4.1 days earlier.

The ripple effect is felt across the plant and animal kingdoms. The scientific field of study is called phenology. It looks at the timing of biological cycles within a range of species as they relate to the climate and seasons. In many species, timing of food sources, migration, or reproduction is linked, forming an intricate interdependence. And across many species there's been a desynchronization, or a decoupling.

An example is the relationship between the miner bee and the early spider-orchid. The orchid, while named after a spider,

---

13 In recent years, these hot spells have not been isolated. February 21st, 2018, wasn't isolated. Temperatures were freakishly warm in 2017 and 2016 as well.

produces flowers that actually resemble a bee, specifically the female miner bee. Since 1848, botanist have kept records of the orchid's blooming time, which coincided with the bee's reproductive cycle. The flowers release a hormone that mimics the female bee and seduces males of the species, which attempt to copulate with the flower. This act of trickery pollinates the orchid. But warming temperatures have thrown this reproductive cycle out of synch. For every 1°C rise in temperature, the orchids bloom six days earlier. The change in temperature has an even more marked effect on the bees, with the males emerging nine days earlier, and the females fifteen. The result is that the male bees are no longer "mating" with the flowers, preferring to mate instead with the female bees. This is good for the bees but not for the orchids, which have come to rely almost exclusively on the miner bees for pollination—and ultimately, existence.

For birds that migrate across continents, the changing flowering and insect cycle is also leading to a timing mismatch. Researchers studying North American avian species have found that some birds—including indigo buntings, northern parulas, great crested flycatchers, blue-winged warblers, and Townsend's warblers—are arriving well after spring starts. The old adage reminds us that the early bird gets the worm, but for some species that are arriving up to fifteen days late in the season, the caterpillars and insects that time their reproductive cycles with blooming plants have already moved on, leaving the birds tired and starving after their long journey.

That's because birds time their migrations with the sun. When the sun begins to rise earlier in the day, this signals it's time for the migration. But the trees and plants are timing their "biological spring" not with light but with a change in temperature. Insects emerge en masse to feed on young leaves before trees release their natural repellent. And by the time the birds flying in from Central and South America arrive for their annual buffet, there's been a fundamental mismatch and a cascading effect. As Stephen Mayor, a lead author of one study, has stated, "The growing mismatch means fewer birds are likely to survive, reproduce and return the following year. These are birds people are used to seeing and hearing in their backyards. . . . It's like *Silent Spring*, but with a more elusive culprit."

The elusive culprit is timing, and around the globe it's throwing many species and ecosystems out of whack. As Damian Carrington writes in *The Guardian*, "Suspected mismatches have occurred between sea birds and fish, such as puffins and herring and guillemots and sand eels. The red admiral butterfly and the stinging nettle, one of its host plants, are also getting out of sync." This is a very real butterfly effect, and its impact could be devastating.

Scientists are also noticing changes at the very base of our own food chain. From honeybees and insects that pollinate the vast majority of our crops to the shifting timing of plankton in the seas, which in turn affects shellfish, fish, seabirds, sharks, and marine mammals like seals and whales, nature's complex systems are under more strain all the time.

Our inventions of the human clock and the manufacturing cycle—that rule our behaviour inside the reality bubble—have begun to wreak havoc with nature's temporal cycle. Not only are we subject to the artificial beat of our own invented time but species throughout the plant and animal kingdoms are starting to feel this rupture as well. The changes around us are accelerating, and yet we have not even noticed that this fundamental break has to do with our own creation of time. Instead, as Bertrand Richard notes, faced with "climate chaos, stock market panics, food scares, pandemic threats, economic crashes, congenital anxiety, existential dread," we do not slow down. Instead we do the exact opposite: we put the pedal to metal and increase our speed.

Before moving on, there is one other clock I should mention that scientists have devised. This one is more metaphorical than physical. Each year since 1947, the *Bulletin of the Atomic Scientists* has used the concept of a clock counting down to midnight to show the imminence of humanity's destruction from climate change, atomic weapons, and other technologies of our own making. It is known as the Doomsday Clock. On January 25, 2018, in an open letter addressed to leaders and citizens of the world, the scientists announced that we are now highly vulnerable to catastrophe, and the clock ticked a minute closer to our end. It now reads two minutes to midnight.

The question before us is: Do we hit the snooze button?

8
SPACE INVADERS

Measurement is certainly an illusion,
because you don't find inches lying around—
you can't pick up an inch. Inches . . . are actually imaginary.
—ALAN WATTS

IN ENGLAND, you can go for a walk through Madonna's private estate. This is possible because the "right to roam" through the countryside is enshrined in British law. In nature, this freedom—for animals at least—is a given. For birds flying overhead and insects marching on the ground, the lines drawn by humans to divide private and public property are largely inconsequential. But for our species, it's like we live in a world that's been booby-trapped with tripwire. As Floyd Rudmin, an expert on community psychology, writes, "more than 99% of the world around us is off-limits to any one of us, and we rarely notice that."

But in the 1930s in England, a group of young people did take notice. They were factory workers in the industrial heart of Manchester. At the time, Manchester was gloomy and polluted, but nearby was the Peak District, a gorgeous part of the country with grassy, rolling hills. The problem was the workers were forbidden from entering it. This was before national parks existed in Britain, so for the workers, going out to take a bit of fresh air and enjoy nature meant they were trespassing on private land.

So on April 24, 1932, a group who called themselves "ramblers" decided to engage in a simple act of protest. They would do something that few of us would consider "rebellious" today: they would go for a hike. But knowing that they'd be stopped if it was only a few of them, the ramblers gathered as a large group instead. Together, four hundred people set off up a mountain in the Peak District called Kinder Scout.

At first, local gamekeepers with clubs attempted to stop them, and a scuffle broke out, but the men backed off when they realized they were outnumbered. Then the police got involved, and several of the hikers were arrested and jailed. Surprisingly, this worked to the ramblers' advantage. The story spread quickly, and public sympathy led to a national outcry as more and more people began to demand the right to walk through the countryside. The Kinder Scout mass trespass is now known as one of the most successful acts of civil disobedience in British history and is celebrated every year. By literally *exercising* their freedoms, the ramblers paved the way for the creation of national parks and opened up nature to the average person.

These days, in England, you can walk through 7 percent of the countryside. That may not sound like much, (and is a reminder of how much land is still off limits), but it's significant enough. In Scotland, however, things are even better; modern ramblers not only have walking rights, they also have access to the equivalent of free Airbnbs. That's because the landscape is dotted with empty farmers' cottages known as bothies, abandoned huts that remain as a relic of the Scottish Clearances, when populations were driven from the rural highlands. These modest huts now form an informal travellers' network of accommodations, and while they are basic—some only have beds and a hearth—bothies have become a beloved Scottish tradition of giving hikers a free place to spend the night.

Freedom sounds wonderful, of course, but it does have a flip side: lack of security. After all, while it might sound lovely to amble unrestricted through the countryside, you might not feel *quite* as comfortable if a stranger were walking through your own backyard. That's because human beings, like many animals, are territorial.[1] It's hard-wired into our brains. Scientists know that animals from insects to chimpanzees have evolved a sense of personal space, and this makes sense, because in the wild having your space encroached upon can be a threat to survival.

For humans, four distinct zones of personal space have been detected. In the 1960s, American anthropologist Edward Hall was the first to measure and define what he called reaction "bubbles." The closest bubble surrounding us is "intimate space." It extends approximately forty-six centimetres around the body and is reserved for family, partners, and close friends. The next bubble is "personal space." This ranges from 0.46 to 1.2 metres and is where we feel most comfort with acquaintances. "Social space" is the third bubble. It extends 1.2 to 3.7 metres from the

---

1 For human beings, space is also emotional. Indigenous groups often speak of the spiritual connection they have to the land. And we've all experienced that feeling with a place we've come to know well. We have fond memories of our favourite places growing up and often a dear attachment to our homes.

body and is the area for strangers and new acquaintances. Finally, beyond that is public space, a space that for the most part anyone is free to enter. There are of course exceptions, and as social creatures things like status, sex, and cultural differences all affect our notions of personal "safe" space, but in general these bubbles define our most basic sense of territory.

The part of the brain responsible for our feelings of fear or safety is the amygdala. It forms the neural circuitry behind our "fight or flight" responses. But in rare instances where there has been damage to the amygdala, the human sense of these spatial boundaries can be erased. This was the case with SM, a patient with significant amygdala lesions who, as a consequence, had no sense of personal space. Like the character Aaron, a notorious "close talker" from an episode of *Seinfeld*, SM was perfectly comfortable standing even nose-to-nose with strangers.

Our need for space, then, at least at close quarters, is an evolved defence. But in defending larger territories—spaces beyond personal threat—there are allowances. In the animal kingdom, birds like the American robin, for instance, are aggressive towards other robins that enter their territory but let in birds like the white-breasted nuthatch, as the two species have different food sources and do not compete. Chimpanzees likewise guard their space from their own species. And their territories are significant, ranging from 48 to 241 kilometres. Males, and sometimes females and juveniles, regularly patrol their borders to ensure that their chimp neighbours do not sneak into their territory. And while neighbouring males are always attacked as intruders, fertile females are welcomed as new immigrants, at least by the males chimpanzees. They are usually fought off by the resident females at first but eventually are allowed in.

Humans, however, are distinct in that we manage our territories not just physically but also with our minds. As a species, we've created borders and maps that separate us from one another and mark out defined spaces. Also setting us apart is that our territories can be massive; in fact, what constitutes our territory now spans the entire globe.

IN FINLAND, THE INDIGENOUS SAMI PEOPLE have a very special unit of measurement. It's called a *poronkusema*, which is defined as the distance a reindeer can travel before it needs to stop and urinate. The Sami, who have lived alongside reindeer for centuries, attentively noted that the animals won't walk and relieve themselves at the same time. And so, once approximately every 7.5 kilometres, a *poronkusema*, they stop and empty their bladders. While this measurement may seem a touch absurd to non–reindeer herders, it should be said that before the metric system came along, many countries and cultures had their own rather peculiar systems. It's likely that people of the future will find it just as weird that we described the unfathomable loss of our rainforests in terms of "football fields."

Our ability to define and measure space accurately separates us from other species. We are the only animals with the ability to project our mental delineation of the world—using lengths, widths, lines, and maps—to define the physical space around us. To understand how we've constructed this mapped world, however, it's important that we first look at how we came to create the basic units of measurement.

The dimension of space, as we saw in Chapter One, is too gargantuan for our little brains to fathom, and so we've scaled down the abyss by chopping it into more manageable pieces we can understand, something human-sized. Much like the dimension of time, originally all measures were made with bodies. As Witold Kula writes in *Measures and Men*, throughout most of history the human body acted as "the measure of all things." Going as far back as 2700 BC, the Egyptians were using the royal cubit, a length that ranged from 523.5 to 529.2 millimetres and was the rough equivalent of the measure from the tip of an outstretched hand to the elbow. The cubit was then further broken down into seven "palms," which were approximately seventy-five millimetres each, and palms were subdivided into four *djeba*, or fingers, of about nineteen millimetres each.

These forearm measures were popular around the world. And it made sense. We all have arms, so it was like carrying your own measuring stick. The ancient Greeks measured the cubit at 460 millimetres, while the ancient Roman *ulna* was 444 millimetres. Certainly not exact measurements, but similar enough. Other body part measures were used, including the *shaku* (foot) in Japan, the *hasta* (forearm) in India, the *chi* (hand) in China, the *thnang dai (finger joint)* in Cambodia, and the *wa (outstretched arms)* in Thailand.

But the problem with measures based on the body, as you're no doubt aware, is that they aren't equal. So when rulers wanted to levy taxes on a population, the exact meaning of one handful or basketful of wheat or sixty strides of a field was subject to interpretation. Farmers, of course, preferred long strides, just as tax collectors favoured big baskets. In ruled territories, then, as sociologist Zygmunt Bauman suggests, all these colourful varieties of measurement had to be subsumed by the "imposition of standard, binding measures of distance, surface or volume, while forbidding all other local, group- or individual-based renditions." Measures had to become standardized.

In Medieval England, land was once measured by the amount needed to support a household. This unit of measure was known as a "hide." Traditionally, a hide was about 120 acres, but the definition of a hide was flexible in the sense that it was more a measure of value (in the form of taxes it produced from the family) than it was a measure of area. A hide of fertile soil, for example, would be smaller than one with poor soil. The point is, measurement of space was not something written in stone. It could be negotiated.

In contrast, the first *standard* that we know of was issued by Richard I in 1196. In the Assize of Measures, the king decreed that "throughout the realm there shall be the same yard of the same size and it should be of iron." But he was also aware that to be accepted, his standards had to be seen as beneficial for the people. And so, when Magna Carta was signed in 1215, it not only set limits on the monarchy and gave greater rights to the rebel barons to gain their political support, it is also, rather surprisingly, delineated the first standards for beer. The charter's "rights" ensured that one's beer would finally be the same "throughout the kingdom," ensuring that townsfolk weren't short-changed by each other or by greedy merchants.

As the charter's thirty-fifth clause reads, "There shall be standard measures of wine, ale, and corn (the London quarter), throughout the kingdom. There shall also be a standard width of dyed cloth, russet, and haberject, namely two ells within the selvedges. Weights are to be standardised similarly."

Over the centuries, as different monarchs came into power, measurements often changed with their desires. During the reign of Edward I (1272 to 1307), the rod, or perch, was the official land measure. Its definition was much like a Monty Python skit, a rod being equal to "the total length of the left feet of the first sixteen men to leave church on Sunday morning." Until the reign of Henry VII (1485 to 1509), the "yard" was proudly defined as the "breadth of the chest of the Saxon race." Henry then replaced it with the standardized "ell," a measure that was approximately a yard and a quarter, a measure borrowed from Parisian drapers. The ell was then superseded by Elizabeth I's yard in 1588. Her yard was fairly long-lived, lasting over two hundred years. But in 1824, another yard took its place when George IV commissioned the Royal Society to come up with the imperial standard. Unfortunately, his yard only lasted nine years and 198 days, as the official rod that measured it was damaged in a massive fire on October 16, 1834, that burned the Houses of Parliament to the ground.

Measurements have not only been subject to the whims of history, they have also differed from country to country, and even from place to place within a country. As Ken Alders writes in *The Measure of All Things*, in pre-revolutionary France, it was estimated "that under the cover of some eight hundred names, the Ancien Régime of France employed a staggering 250,000 different units of weights and measures." It was also the French, however, who came up with the first "universal" standards of measurement, setting up the metric system that most of us use today.

When the French Revolution crushed the monarchy, it also dropped the guillotine down on antiquated thinking. Gone was the old hodgepodge of measures. The French set up a new system that they declared would "be for all people, for all time." Their big idea was to remove the human body—and the complexities of the human form—as the primary site of perspective and instead base new units of measurement on something more universal: our planetary body. The endeavour was huge.

In the summer of 1792, two astronomers, Jean-Baptiste Delambre and Pierre Méchain, set off from Paris in opposite directions. Delambre headed north, while his colleague Méchain headed south. They had an incredibly ambitious goal: to be the first men to measure the world. To accomplish this, they would survey the meridian arc from Dunkirk to Barcelona, which passed through Paris. Extrapolating from that measurement, they would then find the distance from the equator to the North Pole.

It was, and remains to this day, an astounding feat. Using platinum rods to survey the line of longitude, the pair calculated that the distance from the North Pole to the equator was ten million metres, and hailed that the metre would be defined as one ten-millionth of the distance between the pole and equator. Based on satellite measurements, we now know that the exact distance is 10,002,290 metres. Delambre and Méchain were off the mark only by a couple of kilometres, meaning their calculation of a metre was accurate to within 0.2 millimetres, or about the width of two strands of human hair.

This measure of Earth was forged in platinum, and on June 22, 1799, a prototype of this metre, called the "mètre des Archives," was placed in the National Archives in Paris. Soon, copies of the metre would be exported to other countries for their use. But a complication arose. Copies could be scuffed and were prone to wear and tear. So instead, a new metre, called the international prototype metre (IPM), was born. This platinum-iridium bar would instead reflect a "line standard." On it were two marks giving the measure of one metre, avoiding the problem of having the metre damaged at the ends. The IPM would now be housed at the newly instituted Bureau International des Poids et Mesures, just outside Paris, in Sèvres. This became the "official" metre and would soon be copied and used by more than thirty countries.

As you may recall from the fire at the British Houses of Parliament, however, there was still a problem with a physical representation of an abstract thing. The "real" metre (that represented all the others) could still be damaged. And so the platinum metre, as the standard for the metric system, needed to be protected. In Sèvres, a complex system of fire alarms and anti-theft devices guarded the metre. In later czarist Russia, official measurements were likewise kept secure in the Peter and Paul Fortress in St. Petersburg. But even these elaborate precautions were not secure enough. As Witold Kula writes, "The thought that someday, through an earthquake or a calamitous fire, the world might be 'without the meter' was indeed a nightmare. The new regulations, introduced in 1961, have done away with the very concept of 'standard.' Today, the true or invariable meter is defined as 'a length equal to 1,650,763.73 wavelengths of the orange light emitted by the krypton atom of mass 86 in vacuo' and it is reproducible the world over in any properly equipped scientific laboratory."

This was the birth of the transfer standard. Instead of measuring a dimension by its humanly observable quantities, the metre became an abstract, invisible, and untouchable thing. As with the measure of time, we became removed from perceiving the very measures of our own making. Measures are now so exact, they can only be achieved with advanced technology. Even the krypton measure is already out of date. Today, the dematerialized metre has yet a another definition. With the invention of the laser, space is now not just a measure of light but also a measure of time. Stabilized using molecular iodine, the helium-neon laser defines the twenty-first-century metre, as "the length of the path travelled by light in vacuum during a time interval of 1/299 792 458 of a second."

This all seems highly complex for something that should be so simple. After all, as anyone who's shopped at IKEA knows, most of us still measure objects with our hands and feet. But that's the thing about measurement: it defines everything about our world and yet we don't think about it or its origins at all. Quietly, however, measurements shape the very system we live in. As Ken Alders writes, "Measurement is one of our most ordinary actions. We speak its language whenever we exchange precise information or trade objects with exactitude. This very ubiquity, however, makes measurement *invisible*. To do their job, standards must operate as a set of shared assumptions, the unexamined background against which we strike agreements and make distinctions. So it is not surprising that we take measurement for granted and consider it banal."

This banality is formalized into the accepted "way things are."

The boundless world becomes a measured world. Our personal relationship to space changes. Space becomes a thing. It becomes a blind spot. But we don't question the abstraction. And so, over time, the way we structure space begins to seem natural or inevitable to us. But as we shall see next, measurement not only defines the boundaries of our world, it also defines who occupies it.

TODAY, THE U.S. ARMY trains on battlefields known as synthetic training environments (STEs). Already, the army has developed virtual versions of North Korea, South Korea, New York, San Francisco, and Las Vegas as simulated training grounds, full 3-D replicas generated from real environments. The ability to make high-precision measurements is vital, because they flip reality inside out. Here, the map unfolds and becomes the territory.

The idea is that army units can know a battle terrain well in advance should they ever need to set foot on the ground. And while the technology has improved, it isn't new. Already in a 1993 *Wired* magazine article, science fiction author Bruce Sterling wrote of how the military was using virtual space to conquer its real counterpart:

> Project 2851 is about the virtual reproduction and archiving of the entire planet. Simulator technology has reached a point today in which satellite photographs can be transformed automatically into 3-D virtual landscapes. These landscapes can be stored in databases, then used as highly accurate training grounds for tanks, aircraft, helicopters, SEALS, Delta Force commandos. What does this mean? It means that soon there will be no such thing as "unknown territory" for the United States military. In the future . . . the United States military will know the entire planet just like the back of its hand. It will know other countries better than they know themselves.

If the idea of having your city virtually mapped by the Pentagon makes you a little uneasy, you may want to spare a moment for those who, over the centuries, saw Europeans arriving with astrolabes and transits and other cartographer's tools. Whether they were Spaniards in Peru, Frenchmen on the St. Lawrence, Englishmen in Africa, or Mason and Dixon themselves traversing the United States, agents of foreign powers inevitably arrived following the making of their maps, and violence often ensued.

But cartography is also used to maintain peace and reconcile competing interests in a civilized manner. Today, the Arctic—with US$35 trillion worth of oil, as well as fishing and mineral wealth, entombed beneath the ice and water—has become a point of strategic interest. Five nations—the United States, Canada, Norway, Denmark (Greenland), and Russia—have territorial claims that reach into the polar region. So who gets what?

In maritime law, each country can claim two hundred nautical miles (370.4 kilometres) beyond their shore as an "exclusive economic zone." Underwater, however, there is also the continental shelf, which can "belong" to one nation or another. And so, in cases where the continental margin extends beyond the reach of the law of the seas, nations can further claim another "outer limit," which is defined as either: "(i) points 60 nautical miles [111 kilometres] from the foot of the continental slope; or (ii) points at which the thickness of sedimentary rocks is at least 1% of the shortest distance from the points in question to the foot of the continental slope."

It gets complicated and technical, and it is here that the lines that define sovereignty begin to get muddied. That's because wherever a continental shelf extends farther than the 200 miles (370.4 kilometres), a country can tack on an additional 350 miles (648 kilometres) from the baseline (or low water mark on shore) or 100 miles (185 kilometres) from where the shelf reaches a 2,500-metre depth. And so Canada filed a submission to the UN to re-evaluate where its continental shelf ends. Russia too has filed a submission, which would overlap with Denmark's and likely Canada's submission as well. In the meantime, Denmark and the government of Greenland have also proposed a new "outer limit," and that one would overlap with Norway's continental shelf. The lines will tangle over each other like unspooled thread.

For now, though, while the Arctic ice holds, the friction between nations over resources has yet to heat up. Russia did, however, make a symbolic claim to the North Pole in 2007. In a move that sparked international controversy, Russian scientists travelled 4,300 metres deep below the ice in the Mir 1 and Mir 2 submarines. Their goal was to collect sediment and water samples to scientifically prove the continuity of the seabed as part of the Russian continental shelf. They also planted a one-metre-tall titanium Russian flag on the Lomonosov Ridge, where it still stands in the dark waters under the polar ice.

But here's the thing: maps can be wrong even when they're accurate, because what they really assert is not the bounds of space, but the reach of power. And while geology and hard science were used to justify the Russian claim to the North Pole, for the international community it was not enough to make a convincing case. Interviewed for *The Guardian*, Kim Holmén, the research director of the Norwegian Polar Institute, rebuked Russia's claim by saying, "The United States and Europe were at one time connected, the Appalachians and the Scottish mountains are the same geological formation, but Scotland cannot claim the United States is part of its territory because of that. These samples cannot prove once and for all that the whole discussion is over."

Canada's then foreign minister, Peter MacKay, likewise scoffed at Russia's show of planting a flag, stating, "This isn't the 15th century. You can't go around the world and just plant flags and say: 'We're claiming this territory.'"

All of which is true. The continents were once all joined, indisputably. And sticking flags in the ground has not been considered good diplomacy for centuries. But what everyone missed was that the seemingly intractable problem of Arctic sovereignty was not caused by Russian scheming. It was caused by maps. The lines and boundaries that we draw to form nations are arbitrary. That's not to say that continental shelves are imaginary; they're quite real.[2] But the idea that they have anything to do with what we call nations is just that, an idea. And the idea that maps confer ownership begins to look pretty ridiculous when two perfectly accurate maps seem to say two different things. In other words, the point isn't really that one of the maps might be wrong. It's that, as a means to settle the question of who owns the rights to the resources under the ice, they can both be wrong.

ON JULY 24, 1969, customs officer Ernest Murai processed three special arrivals through Honolulu, Hawaii's port of entry. The travellers had been out of the country for eight days, which wasn't particularly unusual. What was unusual was their point of departure. The flight was Apollo 11, and neatly typed on the customs form in the space for the travellers' place and country of departure, it simply read "Moon."

The idea of astronauts returning through customs almost seems like performance art. Neil Armstrong, Buzz Aldrin, and Michael Collins obviously did not need visas or passports when they travelled to the moon, but the moment they returned to Earth they needed stamped pieces of paper to enter. The same

2 With apologies to Bishop George Berkeley.

is true today on the International Space Station. These space travellers circle the planet freely sixteen times a day, but once they land, NASA must bring them their passports so that they have the right to travel again on Earth.

When you think about it, it's remarkable how much power these paper booklets represent, especially considering they are such a recent invention. Today, aside from serving as a form of official identity, they reveal who our allies (and enemies) are on the planet. The most powerful passports grant the greatest freedom, issued by nations that have alliances with many other nations. Individuals who hold passports from Singapore or South Korea, for example, have visa-free access to 163 countries. A passport holder from Afghanistan, however, has access to only 26.

This imbalance of freedom is not something that people with "good" passports think about. As journalist Kanishk Tharoor writes, "Citizens of Western countries like the US are rarely aware of the enormous luxury of their travel documents. Borders melt at the wafting of an American or British passport; the worst inconvenience is often having to stand in line at the airport to collect a visa." By contrast, as Tharoor notes, Syrian refugees travelling to Europe journeyed by boat from Turkey to Greece, then had to hike overland on foot through the Balkans into Central Europe. And it's an expensive journey. A one-way trip of this kind from Syria to Europe costs at least $3,000. But as a Syrian, you can't hop on an airplane and get a visa processed on arrival.

Though passports have technically existed in the form of "travelling papers" for centuries (letters from a king promising safe passage date all the way back to 450 BC), the modern passport as we know it dates back only to 1914. In his book *Closed Borders*, Alan Dowty suggests that this is because until the late nineteenth century the necessary infrastructure simply wasn't in place. "Few governments" he writes, "actually had comprehensive physical control of their borders; nor did they have bureaucracies sophisticated enough to pick out legal from illegal migrants as they passed through border posts."

It was only after the First World War, as empires fractured into smaller nations, that for purposes of security and controlled emigration, the passport came into use. It was a time when countries were far more concerned with people *leaving* their borders than they were with people coming in. As the writer Stefan Zweig recalls in his autobiography, *The World of Yesterday*, "Before 1914, the earth had belonged to all. . . . People went where they wished and stayed as long as they pleased. There were no permits, no visas, and it always gives me pleasure to astonish the young by telling them that before 1914 I travelled from Europe to India and America without a passport and without ever having seen one."

We did not however, only invent the passport. Before the eighteenth century, the world had no nation-states either. Though they seem permanent at any given moment, borders have shifted dramatically over the centuries. If you were to look at them over the period of a thousand years using time-lapse photography, you'd see them writhing across the land like sine waves, particularly in Europe. And though we tend to assume that borders have the important job of neatly separating whatever is on one side of the line from the other, neighbours in the towns of Baarle-Hertog and Baarle-Nassau, for example, which belong to Belgium and the Netherlands respectively, see things rather differently. Here, the borders are about as well defined as a plate of scrambled eggs, thanks to medieval dukes and lords who traded parcels of land back in the day much like currency. As a result, today within one community there are twenty-two mini-Belgiums that serve as enclaves within the Netherlands, and seven of these parcels of land house mini-Netherlands within them. Which means there are "parts of the Netherlands inside parts of Belgium that are inside the Netherlands."

The town has borders criss-crossing it everywhere. Some borders cut right through bars and restaurants, while others split parks and streets and even residential buildings. Some families have split-nationality homes, with a kitchen in one country and a living room in another. Along borderlines, next-door neighbours will have different cable service providers and different garbage collectors. But that's not all. "There are two civic governments—which means there are two elections for two mayors. There are also two sets of regional and national elections. There are two postal services. If you mail a letter from one country to another (which, in this case, means across the street), the letter will take a long route out of Baarle to Amsterdam or Brussels before returning to Baarle. . . . And there are two income-tax rates, two electrical systems, two phone systems, two school systems, and two tennis clubs."

For a time, the bars on the Dutch side of town closed earlier, and for restaurants that were split in the middle, you could bend the rules by simply ushering patrons over to the Belgian side to continue to eat and drink. But this big tangle of dividing lines also created financial loopholes. And while residents still must obey the borders, there are ways to make the system work for them. Taxes, for instance, are paid to the country in which the front door of the house is located, and so in several cases shopkeepers have moved their front entrance, and thus moved to a new country, in order to pay more favourable rates.

All this is to say that the invisible lines we draw are powerful, and while they may not always separate people or cultures, they do separate laws.

In the case of Baarle-Nassau and Baarle-Hertog, clearly the residents have more in common with each other than they do with any other group on the planet, despite the fact that they live in different countries. At the same time, while one nation may claim to be distinct from another, contained within its boundaries there are always ethnic groupings, political factions, and religious differences no matter how seemingly homogeneous the country. That's because there is a tendency to conflate nation-states with territorial states. As John A. Agnew writes in *Mastering Space*, "This seems innocent enough, but it endows the territorial state with the legitimacy of representing and expressing the 'character' or 'will' of the nation. . . . Many states are clearly not nations in this singular sense."

That certainly became the case for many countries in Africa as the colonial European powers carved up the continent without much thought for the people and the ethnic groups who inhabited it. In the Horn of Africa, Somali peoples had their homeland split into British Somaliland, Italian Somaliland, French Somaliland, the Somali region of Ethiopia, and the Somali region of northern Kenya. For nomadic groups and pastoralists in particular, colonial borders restricting their movement often meant that their traditional lifestyles were disrupted if not destroyed. Many had to settle and compete with bordering groups for resources, leading to an increase in conflict.

In other instances, it was the reverse that led to problems: completely separate ethnic groups lumped together within boundaries that did not share customs or a heritage at all. Take one of Africa's newer countries as an example. Angola comprises ten different ethnic groups, who really have only one thing in common: they were all colonized by the Portuguese and in 1975 they overthrew them. The Europeans took what they could with them when they pulled out, but they left their borders and a country that hadn't previously existed.

Basically, belonging to the same country as someone else does not mean you're the same, any more than belonging to a different country means you're different. Think of it this way:

There are approximately 6,500 languages spoken on Earth. And language is one of the primary unifiers of culture.[3] Given that there are 195 counties, that would mean that on average a country contains thirty-three different linguistic cultures. Consider Papua New Guinea, with more than 800 spoken languages, or Indonesia, with 742. Clearly, national boundaries do not map onto cultural groupings. It just doesn't add up.

There is another way of looking at borders. While national holidays, foods, and customs celebrate the things we share with people on our side of the border, it may well be that there are people we want to keep out. We are animals after all, and as animals we sometimes fight over territory. Our animal natures can however be placated when space and resources are plentiful. This is why uncharted space, like outer space, is largely uncontested. There is so much room in the solar system and the universe that it's really pointless for us to fight over it. Similarly, in desert lands, small populations of nomads had the right to roam across vast tracts of territory for millennia. As Fred Pearce writes in *The Land Grabbers*, the changes we are seeing now are relatively recent, as only "a generation ago, the Bedouin and their camels roamed the deserts of the Middle East." But even then, "It wasn't a free-for-all. Rights of ownership and access were tightly negotiated and policed, but without fences, formal laws, or national boundaries."

Because nomads often adapted to harsh and infertile places like the desert, steppe, Arctic, or tundra, these populations were typically small and nimble, taking advantage of vegetation and

---

3 Empires require ideological consonance to maintain rule over vast landscapes. Mao understood this well when in 1949 he instituted Mandarin as the "official" language of China. When people say "Do you speak Chinese?" they are conflating at least eight different linguistic groups and hundreds of different dialects.

game as the people migrated with animals and the seasons. In fertile regions, however, human tribes began to settle into permanent homes. Ten thousand years ago, with the advent of agriculture, we were able to create large food surpluses for the first time. Along with developed settlements that needed to be delivered, more food also meant more people. Clashes occurred when space was at a premium. And when settled groups could no longer expand outwards, a new type of human group formation evolved: we began to expand upward. This was the invention of hierarchy.

Hierarchy allowed for social complexity and the coordinated rule of larger groups. In *New Scientist* magazine, Debora MacKenzie examines the evolution of nations, noting that "larger hierarchies not only won more wars but also fed more people through economies of scale, which enabled technical and social innovations such as irrigation, food storage, record-keeping and a unifying religion. Cities, kingdoms and empires followed." But importantly, "These were not nation states. A conquered city or region could be subsumed into an empire regardless of its inhabitants' 'national' identity." In other words, empires tended to be multicultural thousands of years ago, and that never really changed.

But the real scourge of early city-states and empires was not so much other city-states and empires as it was those who had *not* settled and had not devoted their energy to farming. That is, the free people: the nomads.

From the settlers' perspective, a nomad is someone who doesn't care about your border or your division of space. Those that migrated tended to be physically much healthier and stronger than farmers. Some rode on horseback and used bows for hunting, and as such they were considered natural warriors. So farmers had every reason to fear nomads. History, as we know, is filled with the bloody conflicts of the two groups, from

Biblical times up to the present day. But while stereotypes abound, were nomads really predators, or were they prey?

There are two accounts of nomadic peoples. One suggests they were invaders, extorting and raiding agricultural societies, and another suggests they were defending their traditional ways against expansionist agricultural societies. In truth, both are accurate; it just depends on who you are talking to.

The last indigenous people of the EU, the Sami people, support themselves by fishing as well as herding reindeer. And yet the Sami need to fight the nation-states that exist on their traditional lands. As one Sami man declared, "The governments of Finland and Norway are trying to make salmon fishing illegal for the Sami and give new fishing rights to rich people who have built cabins on our homeland." So who should the rights belong to? People who have lived there for thousands of years but do not "own" the land, or people who may not have any real ties to the land but own the private property?

The question will become ever more pertinent with global population growth putting space and resources increasingly at a premium. Now that we live in a world of states, we consider it a sin to be stateless. Still, the idea of an imaginary line that someone could step over to pass from one state to another is a fairly recent one in human history. It was not until 1648, when the Treaty of Westphalia was signed, that Europe established the nascent concept of what we now call sovereignty and put an end to centuries of violence. Emperor Ferdinand II had wanted to impose Roman Catholicism in his domains (where local powers were increasingly Protestant). This triggered the highly destructive Thirty Years' War, which soon involved nearly every country in Europe.

The bloody conflict lasted from 1618 to 1648, by which time Ferdinand had been dead for eleven years, and resulted in eight million deaths. The 224 principalities, duchies, and other small domains that made up what we now know as Germany (where most of the fighting took place), lost 20 to 30 percent of their population by some estimates, making it one of the deadliest wars in history. In the end, all parties had a strong incentive to sign the peace treaty just to end the nightmare. But it's unlikely anyone knew they would be setting a precedent that would change the world. The novel idea finally agreed to, and which paved the way for the United Nations centuries later, was that state boundaries gave the new territories rights to their own policies and religious practices. It was more or less the right to be left alone, in exchange for the commitment to leave others alone. Today, that's basically what we mean when we talk about sovereignty.

That radical idea had implications at the individual level. Now that political control was no longer exercised in a direct line putatively through the will of God, acting via the king, down to the lowly subject of the state, a new alignment was required for statehood. And this alignment was through investing directly *in the state* in the form of private property. The individual now entered into a contract and became legally accountable for the land that they possessed, and in turn had a vested interest in the larger political entity they belonged to. The territorial state then provided the role of protectorate. As John Agnew suggests, the state became a "harmonizer of society."

For the first time in history, the individual—the commoner—held an important place in society. As Brian Nelson, author of *The Making of the Modern State*, writes, "Individual choice, particularly economic choice, now becomes possible." And with this interdependence between the state and individual, the abstract notion of "rights" came to the fore. After all, "One's claim to the right of property . . . becomes valid only to the extent that one recognizes others' claims and accepts the legal duty to accept them." And while the notion of private property existed as far back as the ancient Greeks, for the Greeks, property wasn't something an individual had the right to trade in. Land transfers between families were difficult and required religious authorization, and private land appropriation by the state was unheard of except in cases of exile. Only with the creation of the modern state did we come to see an allegiance to abstract space. The common person's idea of "my country," "my state," and "my property" could only evolve with the idea that the space in question could be thought of as "mine" and bought and sold as a "thing."

YOU MAY NOT SEE the divisions of "space," but they are everywhere. Even the column of space above your head and below your feet has rights attached to it. Home ownership is one way to own space, but the air rights above and the subsurface rights below might belong to somebody else. This was not always the case. For most of history, if you held property it was assumed that what was above and below it was also yours; the space that "belonged" to you extended into infinity. The idea was based on a doctrine called *ad coelum*, which can be traced back to the thirteenth century. In essence, it declared, "*Cuius est solum, eius est usque ad coelum et ad inferos*," which is Latin for "Whoever's is the soil, it is theirs all the way to heaven and all the way to hell." Pretty dramatic for real estate.

The idea of owning space up to the heavenly kingdom was brought back down to earth in a rather unlikely way. It was dismantled not by soldiers with swords or guns but by hot-air balloons and an angry chicken farmer. In 1783, after the first hot-air balloon had slowly drifted across the French countryside, the question of air trespass arose, and soon, with the advent of air travel by plane the question was pushed to the fore. Once that had happened, it was only a matter of time before restrictions came into force. In 1925, the United States Congress passed the Kelly Act, which granted "feeder" routes to airmail services. Then in 1926, the Air Commerce Act established official airways that granted the government air rights to all airspace above 500 feet (152 metres) in urban areas. This meant that trespass laws could no longer apply.

But not even the United States Air Force could violate the sanctity of private space. In 1946, a plucky chicken farmer named Thomas Lee Causby, from Greensboro, North Carolina, took the government to court for aerial trespass over his land. Causby's farm was less than half a mile from Lindley Field, an airfield used by the military in the Second World War. Each day, as the planes took off and landed, they would fly just twenty-five metres above his chicken coops. The blistering roar of the engines put the chickens in a state of shock and they stopped laying eggs. As the case file notes, "As many as six to ten . . . chickens were killed in one day by flying into the walls from fright. The total chickens lost in that manner was about 150." Causby was forced to shut down his business.

In court, Causby won, and for his suffering (and the birds'), was compensated for flyovers between 83 feet (25 metres) and 365 feet (111 metres), the first height being the lowest level at which airplanes flew over his property, and the second height establishing the zone of public easement in rural areas. And while the ruling may have been relegated to dusty law books, it is remembered for one thing: it brought *ad coelum* to an end. The government could not claim that it "possessed" the space down to ground level any more than land owners could extend their property rights indefinitely up into the sky.[4]

With the recent advent of drones however, the law has had

---

4 As the Supreme Court ruled, "The common law doctrine that ownership of land extends to the periphery of the universe has no place in the modern world."

to make additional adjustments to air property rights. The Federal Aviation Administration does not allow unmanned aerial vehicles to fly higher than 400 feet (122 metres) so that they do not interfere with low flying aircraft. But with the ability to hover over homes and peer into windows of residential towers, drones and property rights still clash in this legal grey area.

That's what happened on July 26, 2015, in Bullitt County, Kentucky, when David Boggs watched in horror as William Merideth aimed a shotgun at his new Phantom quadcopter and with one loud bang brought it down. Merideth claimed the drone was trespassing over his property, and while the two men disagreed on the height of the flying machine—Boggs said it was two hundred feet (sixty-one metres) while Meredith said it was under one hundred (thirty metres)—the craft was still within the airspace of the private landowner. And so, according to the judge, Meredith was within his rights to shoot the drone down.

Zooming farther up, even more airspace has been contested, measured, and divvied up. As satellites orbit our planet, they fly not just over individual properties but entire nations. So countries have had to agree on the boundaries where they give up their sovereign air rights. And while there remains some disagreement, in general the horizontal band from 30 kilometres (where the highest planes and air balloons fly) up to 160 kilometres (where the lowest orbiting satellites pass) is designated as the zone where a nation loses its claim to air rights.

Within this band is another invisible border. According to the Fédération aéronautique internationale, which serves as the governing body for astronautics and aeronautics, the official line between Earth and outer space is at one hundred kilometres. Beyond this limit, any human traveller is considered an astronaut. This is also known as the Kármán line, where our planet's gravity begins to lose its grip on spacecraft.[5] Once this line is crossed, a set of laws applies that is different from those on Earth, as defined by the United Nation's 1967 Outer Space Treaty. In space, no nation can claim territorial sovereignty. It is ironic however, that it is out there and not here on Earth that space is free and a "province for all mankind." That the shared common exists in a place that most of us can't access, let alone inhabit.

Peace rules (at least for now) in space, and weapons of mass destruction are strictly forbidden. As a domain that is open to all for exploration, claims of ownership over title or territory may be prohibited, but that doesn't mean that there aren't scams. There are companies online for instance, that offer the rights to "purchase" a star's name, but none of them are officially recognized. The International Astronomical Union (IAU) is currently the only governing body empowered to name space objects, and "as an international scientific organisation, the IAU dissociates itself entirely from the commercial practice of 'selling' fictitious star names, surface feature names, or real estate on other planets or moons in the Solar System."

While it is true that individuals cannot make claim to celestial bodies, private companies are laying the groundwork to exploit resources in outer space. The United States Congress recently passed the Space Act of 2015, which has led some scholars to argue that it violates the Outer Space Treaty. With advances in privately funded space flight and in the spirit of capitalism, the act encourages competitiveness. It has been updated to allow US citizens to "engage in the commercial exploration and exploitation of 'space resources,'" including water and minerals, opening the door to potentially mining comets and asteroids

---

5 Harvard astronomer Jonathan McDowell has argued the Kármán line should in fact be set at eighty kilometres, as some elliptical satellites have been observed at this altitude without "crashing to Earth."

for metals like gold, silver, iridium, osmium, palladium, and platinum—just as Americans once sailed the seas looking for guano.

All of us exist on a tiny speck within our galaxy. That being the case, our claims to owning real estate in outer space are about as absurd as an ant claiming to own all of the real estate in New York City. For scientists and entrepreneurs, however, this quest to create settlements in outer space is the new frontier. Given the limited resources on Earth and our growing population, moving to another planet is seen by some as "humanity's hope for a future beyond Earth."

In the meantime, our home planet continues to be exploited. The world beneath our feet holds immense value, as it contains the raw metals that are reshaped into our cars, trains, and planes; the ingredients for the drywall, glass, concrete, and brick we use to build our cities; the minerals that power communication through our laptops and cell phones; and, of course, the soil that grows our food. It should be no surprise, then, that the ground is the biggest battlefield of all.

ON THE CORNER OF SEVENTH AVENUE and Christopher Street, next to a cigar shop, you'll find the smallest piece of real estate in New York City. Shaped like a slice of pizza, the plot of land is a mere five hundred square inches (3,226 square centimetres). It is covered in a mosaic that defiantly reads, "Property of the Hess Estate which has never been dedicated for public purposes." It was the last sliver of land held by David Hess, who refused to give up a fight with the city to keep his private property.

In 1910, New York City began buying land and demolishing buildings in order to widen the streets and build a subway. Hess refused to sell his five-storey building but was forced to relinquish his property anyway under laws of eminent domain.

All he was left with was five hundred square inches. If that weren't enough, the city rubbed salt into the wound when they requested that Hess donate what was left to be used as part of the sidewalk. Hess refused, which is why this tiny piece of "spite" real estate still exists today.

In 1938, Hess's triangle sold for just $2 per square inch (6.5 square centimetres); adjusted for inflation, the triangle's price would now be more than $17,000. Land at that price works out to $106.6 billion per acre, though Hess's cement sidewalk property yields nothing. No one will ever harvest crops there, or prospect for gold, or draw precious water. Cities are the centres of power and capital in the world, and New York is about as central as it gets. So you're not just paying for land; you're paying for scarce land that everyone else wants really badly.

Today, real estate is the business of buying, selling, and trading space. But the idea that land can be "ours"—like measurements, borders, and nation-states—is yet another human invention. Flipping through real estate listings, you might forget that this idea is only a few hundred years old. As Simon Fairlie, editor of *The Land*, writes in the magazine, "'The idea that one man could possess all rights to one stretch of land to the exclusion of everybody else' was outside the comprehension of most tribespeople, or indeed of medieval peasants. The king, or the Lord of the Manor, might have owned an estate in one sense of the word, but the peasant enjoyed all sorts of so-called 'usufructory' rights which enabled him, or her, to graze stock, cut wood or peat, draw water or grow crops, on various plots of land at specified times of year."

These were the "commons." Here, villagers and peasants shared cropland, pasture, and forest areas. And while plots were subdivided to a degree—a peasant might have a small garden by his home, for example, while another had an area where their animals regularly spent time at pasture—land was not "owned" in the way we think of it today. Use and custom meant more than title deed and measurement.

For use of the land, peasants gave lords and landowners a share of the crops, along with their allegiance. But for landowners, this wasn't always enough, as they realized that idleness would result once the peasants had produced enough food to feed themselves and their families. Writing to the editor of a commercial and agricultural magazine, one landowner lamented, "When a labourer becomes possessed of more land than he and his family can cultivate in the evenings . . . the farmer can no longer depend on him for constant work." In other words, people (quite logically) would kick their feet up once they had laboured enough to put food in their mouths.

For landowners however, this behaviour was deemed laziness. Just as we saw in the last chapter, on time, as soon as space became an abstraction and a commodity, the interests of labour and capital diverged: the rich came to hate idleness just as much as the poor hated being overworked. But profit off the land came not only from use of workers' time. A key motivation for landowners to change the system of the commons derived from a new form of profit: England was becoming renowned for its high-quality wool.

From the fourteenth to the seventeenth century, wealthy landowners fenced off open pastures into "enclosures" and kicked out the peasants in order to privatize the land for grazing sheep. With the help of Parliament, these evictions were enforced by law. If 80 percent of land was owned by one title (as it often was with the wealthy landowners) that meant it could be officially enclosed.

By the mid-1800s, this land grab covered about one-sixth the area of England, or seven million acres, and after four thousand acts of Parliament, what was once common land became enclosed. For the peasants, there was little recourse. As a literate commoner wrote to his landlord in 1824, "Should a poor man take one of your sheep from the common, his life would be forfeited by law. But should you take the common from a hundred poor men's sheep, the law gives no redress."

With the profits to be had in the wool market, more land was cleared. From the mid- to late eighteenth century, and continuing into the nineteenth century, cottages were burned and thousands of families forcefully evicted from the Scottish Highlands in what came to be known as the "Clearances." The result was a massive migration, with most Highlanders heading to the Lowlands in search of factory work, while other clans were shipped off to find work and land in the United States or Canada.

This was the beginning of urbanization. The dispossessed who were no longer able to grow their own food often had no other choice but to move to urban centres and become factory workers. By 1760, the Industrial Revolution had begun, and the machines were hungry for cheap labour. Along with these forced evictions, then, was opportunity, which produced a vector for rural flight. In England and Wales in 1801, 65 percent of the population lived in rural areas; exactly one hundred years later, only 23 percent of the population remained.

Enclosures may seem like a relic of the distant past, but nearly identical processes are in effect in poor countries today. According to Oxfam, in the last decade alone, eighty-one million acres of land worldwide (about the size of Germany) has been seized from peasants and rural farmers and sold off to foreign investors.

But it's not always the fault of the foreign buyer. Often they are told the land is "uncultivated," the assumption being it is unoccupied. But as land reform experts have noted, uncultivated means not farmed, which does not mean the land is unowned or unused. In Africa, for instance, as Fred Pearce writes, "About four-fifths of the continent's 6 billion acres is not formally owned by anyone other than the state. There is no legal title, but rural inhabitants regard it as theirs."

As illustrated by David Hess's case, land can also be appropriated through eminent domain, where the government recognizes property ownership but still forces a purchase of land to convert it from private to public space. This has been happening on a large scale in China. Transforming the country from an agricultural backwater into an empire of gleaming metropolises in the space of a few decades has meant the physical displacement of fifty million Chinese farmers since the 1990s in the name of economic development. In the first decade of this century alone, almost a million villages were abandoned or razed. According to research by Tianjin University and China's Ministry of Civil Affairs, between 2000 and 2010 China went from having 3.7 million traditional villages to 2.6 million. That is the equivalent of losing three hundred villages a day.

China's National Bureau of Statistics estimates that by 2034 less than 25 percent of the country's population will be rural. This trend, from periphery to centre, is taking place all around the world as people flood from rural areas into the cities. UN-Habitat has estimated that around the world in 2009 alone, three million people were moving to cities every week, and by 2030 one-third of humanity will be living in urban centres. As the human population grows and more people move into the cities, the available space for us to share shrinks dramatically.

IN PARIS, NEW YORK CITY, and London, in some of the highest priced neighbourhoods in the world, a new real estate phenomenon has emerged: "zombie flats" and "ghost mansions." They are not haunted houses but rather homes that literally have no soul. If they are spooky, it is only because they are empty. In London, seven out of every ten residences in these prime areas are investments for overseas buyers; in Paris, the number is one in four; and in Manhattan, between Fifth Avenue and Park Avenue from Forty-Ninth Street to Seventieth Street, almost a third of homes are unoccupied for ten months of the year. The properties are vacant because for the super-rich these spaces are not their primary residences. They are often third or fourth homes. This type of "land banking" is now being documented in major cities around the world, with the number of empty residences on the rise in cities like Miami, Jerusalem, Hong Kong, Vancouver, Dubai, Singapore, San Francisco, and Sydney.

The super-rich, like the kings and lords of old, own the vast majority of the land. In England, nearly half the country is owned by just 0.06 percent of the population. In the meantime, as Simon Fairlie writes, "Most of the rest of us spend half our working lives paying off the debt on a patch of land barely large enough to accommodate a dwelling and a washing line." Considering the crush of people in urban centres, the imbalance is obvious. In 2018, in the United Kingdom, 216,000 homes were empty for six months or more; at the same time, 78,000 families were in temporary accommodations or homeless. The numbers in the United States were even more egregious after the 2008 financial crisis: for every homeless person there were five empty homes. We may well be the only species on Earth to actively deprive our own kind from inhabiting a space when it is available.

In China, the discrepancy is stark. In his book *Ghost Cities of China*, Wade Shepard writes that "the world's most populated country without a doubt has the world's largest number of empty homes." Rapid development in China during its economic boom outstripped the demand for new homes. In its rush to build, China used more concrete between 2011 and 2013 than was used throughout the twentieth century in all of the United States. Looking at mobile phone usage and internet activity to measure how much life could be detected in these vacant zones, researchers from Peking University in Beijing found that there are approximately fifty "ghost cities" in the country that remain largely uninhabited. Like an architectural model writ large, apartment blocks, shopping malls and plazas, parks and playgrounds have all been built; the only thing is, like a zombie town, there are no people. The cities are empty.

But with massive migration from rural areas to the city, you might be wondering where all those people go. Lured in with promises of work and a chance of achieving wealth, many Chinese migrants come to live in the factories that employ them. As George Knowles writes in the *South China Morning Post*, some of these factories look like labour camps. Dormitory blocks built to house fifty thousand people are filled with steel bunk-beds (some with twelve beds per room) and workers wash in communal showers. For this, they have about ¥160 (US$25) deducted from their salaries a month.

In cramped cities like Hong Kong, where the average 37-square-metre apartment costs US$2,000 a month, the poor have the same average living space as jail prisoners. A survey conducted by one tenants' rights group found that the average space occupied by families living in partitioned cubicles in Kwai Chung, a suburb of Hong Kong, was 4.65 square metres, or "roughly the size of three toilet cubicles or half the size of a standard parking space." Prisoners, according to Hong Kong Correctional Services, have an average space of 4.60 square metres per person. But believe it or not, even that amount of space is a stretch for some. Some subdivided flats in Hong Kong, called "coffin cubicles," are so small they total only 1.4 square metres.

We have become so blind to the fact that our rules of space are artificial, and that we have created a system where we imprison ourselves. Today, the wealthy in our world own empty ghost mansions, while the poor live in coffins.

In big cities on the Chinese mainland, the situation is just as extreme. For the urban poor who live centrally to avoid long commutes, there's a shortage of affordable housing, and as such many have chosen to live underground. In Beijing, an estimated one million people live in a subterranean network of former bomb shelters. They have been called the Rat Tribe. Rents are half of what they might be above ground, costing an average of ¥436 (US$70) a month for an underground room that's just over 9.75 square metres with shared kitchens and bathrooms. The cramped facilities are often unhygienic. In one documented case, the occupants of eighty rented rooms shared a single toilet. These are people who cannot afford something as simple as sunlight. As Zhuang Qiuli, a young pedicurist who lives underground, put it, "There is no difference between me and the people who live in the posh condominium above. We wear the same clothes and have the same hairstyles. The only difference is we cannot see the sun."

9
HUMAN ROBOTS

And what is a good citizen?
Simply one who never says, does or thinks
anything that is unusual.
—H.L. MENCKEN

The Eyes Above Us

IT WAS OBVIOUS that the dot on the map had stopped moving, but it was some time before people realized that the dot on the map had died. That dot was Michael Hall, a cyclist in a 5,500-kilometre endurance race being watched online by a community of "dot watchers," fans who kept track of the cyclists along their thirteen-day route from Fremantle to Sydney, Australia.

Each dot had an athlete's nametag, and it wasn't unusual for them to pause here and there, when the athletes stopped to rest, take meal breaks, or go to the bathroom. The GPS live trackers were on board the bikes to ensure that cyclists didn't cheat, while also providing the fans with live coverage of the riders they were following.

Over the course of the race, people began to warm up to the little dots moving about the screen. As Belinda Hoare, one of the online trackers, said, "You'd go from checking maybe once or twice a day, to checking a couple of times a day, to checking hourly, and then you'd have the map open constantly. . . . You really did feel like you got to know these people."

On March 18, 2017, Hall's dot was in second place when at 6:22 A.M. it suddenly stopped moving near the intersection of the Monaro Highway and Williamsdale Road. It was the final day of the race, and puzzled fans began to wonder why Hall had paused at this critical point. What they did not yet know was that he had been struck by a car and killed. By GPS, they had witnessed his death.

In a few short decades, GPS has become so ubiquitous and indispensable it has entered almost every sphere of our lives. The moment you step outside with your smartphone in hand, you too are a moving dot.[1] And though you cannot see it, the world has been overlaid with a time-and-space grid. All of us are synchronized to it and can be traced by our coordinates. We are largely unaware of the role satellites play in our lives, but stock markets, telecommunications, jogging routes, drone strikes, local weather forecasts, ATM machines, traffic lights, and food deliveries all rely on this public infrastructure that orbits silently high above us.

The signals are controlled by the U.S. Air Force and used by

---

1 If you live in the U.S and Canada, even your front door has been tagged with its own GPS coordinate. The census uses GPS to record the coordinates of all of the nations' home addresses.

about a billion people daily. At night, you might occasionally see one of these GPS satellites twinkling in the sky like an artificial star as it reflects the sun. Each satellite weighs around two metric tons, and with their solar panels stretched out, the largest have wingspans of about thirty-five metres, or the length of two tractor-trailers. Orbiting at an altitude of 20,200 kilometres above Earth, each satellite belongs to a constellation of twenty-four to thirty-one GPS satellites at any given time, that zip around the planet at speeds of more than eleven thousand kilometres an hour.

In their book *GPS Declassified*, Eric Frazier and Richard Easton imagine what Captain Cook, who navigated with *real* stars, might think of this modern technology.

COOK: What use are invisible stars?

COMMANDER: We don't need to see them. The satellites transmit electromagnetic frequencies . . . radio signals that our equipment uses to determine our position. Radio signals are very rapid vibrations that our instruments detect with their antennas, which for them are like our ears, but these are not sounds anyone can hear.

COOK: You steer your ship with sounds you cannot hear from stars you cannot see?

That is exactly what we do. Humans can hear sound at a range of twenty hertz to twenty kilohertz, but GPS radio signals are much higher than that. Operating at bands of 1,227.6 megahertz and 1,575.42 megahertz, these radio waves, when they reach the ground, aren't audible to any animal on Earth. Radio signals are incredibly faint. As Carl Sagan once said, "The total energy picked up by all the radio telescopes on the entire planet in all of history is less than the energy of a single snowflake hitting the ground." That statement was made when he recorded the television show *Cosmos* in 1980. According to astronomer Frank Drake, who made the original calculation, with the additional radio waves beaming down to Earth since then, the amount might now be equal to "two snowflakes . . . maybe three."

Using receivers, however, we are able to pick up the faint waves of these signals as they wash over us. And it's not just the US complement of satellites we can choose from either. Russia has its GLONASS satellites, the EU has Galileo, and China has its BeiDou system of navigation satellites. Pulling in signals from one or more of these systems, along with a base station at a known position for reference, today's civilian GPS receivers can now pinpoint your location within 1.5 metres (it had been within an area about the size of a football field in the 1990s).[2]

Of course, GPS satellites, which operate at medium Earth orbit (MEO) and circle the planet twice a day, are not the only eyes in the sky. At low Earth orbit (LEO), or an altitude of two thousand kilometres and below, you'll find the majority of Earth observation satellites. Circling Earth every ninety minutes, these satellites are close to the planet's surface and are frequently used for meteorology, map-making, and environmental monitoring. Zooming up much higher, to an altitude of 35,786 kilometres is where you'll find the satellites operating at geosynchronous orbit (GSO) and geostationary Earth orbit (GEO). These satellites are synchronized with the rotational

---

2 On May 1, 2000, President Clinton turned off "selective availability" for GPS, effectively giving civilian GPS the same capability that existed for the military. By flipping the switch, he instantly made the signals ten times more accurate than they had been previously. GPS accuracy outdoors is approximately five meters. With the development of WiFi round-trip time technology that can be used indoors however, accuracy can be sharpened to the one to two meter range.

period of the planet[3] and are primarily used for telecommunications, where the signals can be constantly and reliably accessed from the same spot on Earth. As artist and geographer Trevor Paglen points out, geostationary satellites are "thousands of times further away" and "remain locked as man-made moons in perpetual orbit long after their operational lifetimes." Because they are too far away to bring back down and burn up in the atmosphere, instead, when these satellites reach the end of their lives, operators on Earth use on-board propellant to bump them up three hundred kilometres into what's known as a graveyard orbit. Paglen notes that here they will stay circling Earth, outlasting the pyramids as remnants of our civilization. Archaeologists of the future will not just dig the earth; they will likely uncover much about the twenty-first-century human record by examining our well-preserved machines looming high up in our space cemeteries.

Beyond LEO, MEO, and GEO, however, there are still other orbits that for security reasons have no published schedules. These are the secret paths of the spy satellites. The CIA has been launching "Keyhole" (KH) class reconnaissance satellites since the 1960s. They have powerful lenses that can zoom in on the tiniest details on Earth and yet remain invisible to the general public. As was the case with GPS, the capabilities of civilian craft are several years behind what the military can do. In April 2018, Surrey Satellite Technology Ltd. announced that the United Kingdom's Carbonite-2 satellite was able to record full-colour

---

3 Geosynchronous satellites orbit the planet in synch with our sidereal day, or once every twenty-three hours, fifty-six minutes and four seconds, and thus appear to an Earth observer to be in the same place at the same time once a day. A geostationary satellite also orbits the planet once a day but is stationed high above the equator and thus will appear to an Earth observer to be stationary in the same place throughout the day.

HD video from 505 kilometres away. By stacking the frames in much the same way macro photographers do, experts can use the data to resolve an image from space down to sixty centimetres. And this is a civilian craft. With some US government restrictions relaxed, commercial imaging satellites like the ones Google uses will now be able to show images at twenty-five centimetres of resolution. That's the ability to see your face—from space.

The latest spy satellites are even more powerful. Keyhole electro-optical imaging satellites of the KH-12 class are said to have a primary mirror on board that is 2.4 metres in diameter. That's the same size as the mirror used on the Hubble Space Telescope, which is used to image objects ten to fifteen *billion* kilometres away. We can only imagine the resolution this class of reconnaissance satellites has when its lenses are turned not on outer space but on Earth.[4] Not much is revealed about the Keyhole satellites, but we do know that launches are managed by the National Reconnaissance Office with a heavily funded budget of an estimated $10 billion a year. The satellites usually follow polar elliptical orbits, allowing them to scan all of Earth from pole to pole, passing over the equator at a different longitude each time as the planet beneath spins from day to night.

Even declassified documents about spy satellites are heavily redacted, so what we know about their locations comes largely from hobbyists who track their trajectories from Earth. These amateur astronomers have their eyes trained on the top secret machines. They are, in essence, the only eyes that watch the watchers. Communicating through a mailing list called SeeSat-L,

---

4 Imaging resolution is said to be at least twelve centimetres, meaning from space you could see a pair of tweezers on the ground. But given the lens size, and the fact that commercial providers are already pushing for ten-centimetre resolution, Keyhole-class satellites can likely see a resolution much higher than that.

this small group of observers from around the world uses stopwatches, telescopes, and cameras to monitor the orbital planes of approximately four hundred military satellites. As one member of the group, Marco Langbroek, puts it, "Just like the Earth has a coordinate grid, with latitude and longitude, the sky has a coordinate grid, and every star has a coordinate within that grid. And by using stars as a reference point you can determine the coordinates of a satellite in the sky."

For most of us, these satellites are out of sight and out of mind, but the reality is there are thousands of highly advanced commercial, scientific, and military eyes that hover and swoop in the skies above us performing duties that are critical to the functioning of modern society. As geostrategist Nayef Al-Rodhan writes,

> Any accidental interruption or deliberate severance of space-based services would cause immense financial losses and other disruptions. Indeed, a single day without access to space would have disastrous consequences worldwide. Approximately $1.5 trillion worth of financial market transactions per day would be stifled, throwing global markets into disarray. According to statistics provided by the International Air Transport Association, over 100,000 commercial flights crisscross the planet daily. Evidently such flights would be interrupted by communication disruptions, and deliveries of emergency health services would be severely hampered. Additionally, coordinating effective responses to crises would become nearly impossible. Due to the fundamentally transnational nature of almost all outer space activities, any conflict in outer space—even a limited one—would have disastrous consequences for the large amount of civilians globally who depend on the provision of outer space services. Contemporary strategists warn that command and control structures of modern militaries are also becoming

> critically dependent on space-based assets for communication, coordination, reconnaissance, surveillance, high-precision targeting and other critical military activities. This increasing indispensability of space for modern military activities makes satellites ideal targets in future conflicts.

The targeting of satellites makes everyone on Earth—especially the most developed and technologically advanced nations—exceptionally vulnerable. One incident in particular has shown the potential for serious disruption in space. On January 11, 2007, China launched a ballistic missile from Xichang Space Center. Its target was innocuous, the Fengyun-1C, an old Chinese weather satellite travelling at around 27,000 kilometres an hour. The missile carried a kinetic kill vehicle,[5] which it released toward the weather satellite, coming in from the opposite direction at a relative velocity of 32,400 kilometres an hour. The head-on impact destroyed the satellite instantly, shattering it into a cloud of debris that sent over 35,000 shards into orbit, where they still circle like a ring of daggers around the planet. The threat of the space debris to other orbiting satellites is certainly dangerous, but the mission made something else crystal clear. While ostensibly China was decommissioning an aging satellite, it also proved to the world it had the capacity to destroy a satellite in orbit and blind another nation's eyes.

The Eyes Around Us

"TO THE LADY in the brown dress," the voice from the CCTV camera said, "blond hair, with the male in the black suit, could you please pick that cup up and put it in the bin." The speaking camera is one of a network of 144 such surveillance devices in

---

5 The Outer Space Treaty banned weapons of mass destruction in orbit and outer space but does not ban conventional weaponry in orbit.

the town of Middlesbrough, England. There are over twenty towns in England where Big Brother doesn't just watch over you, he barks out orders and literally tells you what to do. For stopping litterbugs, the approach seems harmless. In North London, however, where similar cameras are installed on public housing developments, they are oppressive, especially when people standing outside their own homes are told they are loitering. But this type of talking surveillance is not just in poor or middle-class areas either. In Mandelieu-la-Napoule, one of the French Riviera's wealthiest towns, talking CCTV cameras were installed to reprimand people for infractions including bad parking, not picking up dog poop, littering, and other antisocial behaviours. As the deputy director of *Le Parisien* newspaper wrote, the new system is like "a voice from the heavens to warn you not to step out of line."

All of us are surrounded by cameras today, most of them silent. The United Kingdom, home of George Orwell, has the dubious honour of having the greatest number of surveillance cameras in Europe per capita, with over six million CCTVs, or about one for every ten people. The United Kingdom also utilizes automatic number plate recognition, and with approximately nine thousand cameras, it captures up to forty million pieces of data in the form of number plates per day and is currently holding about twenty billion records. As a report by the UK government's independent Surveillance Camera Commissioner notes, this makes it "one of the largest non military databases in the UK."

China has not surpassed the United Kingdom's number of surveillance cameras per capita, but with its massive population it certainly has the greatest number in operation of any nation. There are more than 170 million CCTV cameras installed in the country, and that number is expected to jump to between four hundred million and six hundred million by 2020.

Networked with artificial intelligence (AI) and facial recognition, the new surveillance hubs in China—with towering floor-to-ceiling digital screens and glowing semicircular command and control desks—are as sleek as anything you might see in a futuristic sci-fi film. In one of these hubs, in the city of Guiyang, the database contains a digital image of every single resident. For local citizens, the networked cameras track a person's face from their ID card and trace their movements back through the city over a timeline of one week. Connecting a person's face to their licence plates, and expanding through their contact list of friends and family, the system also knows "who you are and who you frequently meet." In addition to recognizing individual faces, some systems can estimate age, ethnicity, and gender.

To see how the Guiyang system works, the BBC designed a clever "Where's Waldo?"–type experiment. They set their reporter John Sudworth free in the streets to find out how long it took to track him down. Flagged as a "suspect" for the purposes of the trial, Sudworth was no match for the AI powered eyes. He was tracked down and apprehended in just seven minutes.

But we are not only tracked outdoors. If you walk through any shopping mall, office, or commercial space and look up, you will see the ubiquitous black domes. They are tinted so that the camera can look out but you can't see where the lens is pointed.

Increasingly, our conversations are also recorded by invisible ears. As William G. Staples writes in *Everyday Surveillance*, "Public buses in San Francisco; Athens, Georgia; Baltimore; Eugene, Oregon; Traverse City, Michigan; Hartford, Connecticut; and Columbus, Ohio, have been equipped with sophisticated audio surveillance systems to listen in on the conversations of passengers." And in Las Vegas, Detroit, and Chicago, the Intellistreets system has been installed. These are street lights and lampposts with embedded microphones and cameras that are capable of secretly recording pedestrians' conversations.

At our workplaces, even office cubicles are increasingly surveilled. As an article in *MIT Technology Review* notes, this form of high-tech office surveillance is invisible, because "sensors are hidden in lights, on walls, under desks—anywhere that allows them to measure things like where people are and how much they are talking or moving." All of this, of course, is under the guise of improving productivity and cost savings. Companies like Humanyze provide what they call "people analytics." Workers are given ID badges with embedded microphones, Bluetooth sensors, and accelerometers, and data is quietly collected behind the scenes as people go about their day. The idea is that by tracking where the workers are, who they are talking to, and for how long, management can, for instance, understand which departments have the best information flows and make strategic decisions as a result, even improving communication through floor planning. The system also tells managers how productive people are by analyzing how much time a person spends socializing at work. Eerily, the devices can even track "how long an individual goes without uttering a word to anyone—and when that word does come, where does it happen and to whom is it addressed."

Right now, a full three-quarters of US companies subject their employees to regular workplace surveillance. In many ways, we are revisiting a high-tech remake of the extreme Taylorism that we saw in Chapter Seven, where the minutiae of labour productivity are "guided" by principles of scientific management. The cameras, sensors, and smart systems are in themselves a form of "super vision" for supervision. And with the current video surveillance market worth US$36 billion and projected to reach US$68 billion by 2023, surveillance is constantly being sold to us as a tool that enhances efficiency, safety, and security. But behind these promises there is a darker, less benign force at work. As Staples writes, modern surveillance strategies are "used by both public and private organizations to influence our choices, change our habits, 'keep us in line,' monitor our performance, gather knowledge or evidence about us, assess deviations, and in some cases, exact penalties."

It is not said often enough: what is most eroded in a surveillance society is human trust. Instead of trusting each other, we put that trust in spying eyes, GPS trackers, and networked machines. As employees who work offsite, commercial truckers experience this kind of insidious monitoring on a daily basis. Electronic monitoring has become a high-tech way for managers to watch over their performance. And fleet owners do make a good argument for the case: monitoring results in more frequent seatbelt use, higher productivity, less speeding, less overtime, and lower fuel usage and thus a lower carbon footprint. On paper, this sounds great, but drivers tell a different tale. For them, being constantly tracked by telematics is dehumanizing and oppressive.

Telematics refers to the recording and tracking of long-distance data, primarily vehicle data—which includes mapping and routes, driving speed, idling time, acceleration and braking, and seatbelt usage, to name a few—and is designed to keep workers constantly on task and operating optimally, essentially behaving like human robots. After each shift, the data is uploaded to a computer and then transmitted to a data centre, where it is algorithmically analyzed. While it's good for business, for the drivers to have to explain any deviation from the path or time misspent is humiliating. As one driver put it, telematics "should be known as Harassamatics." Another stated that the data made him look guilty when he was innocent: "They assume that every driver is cheating and stealing from the company, they just haven't caught them yet. Telematics brings a whole new perspective to this world of assumption. . . . Every time Telematics has been thrown in my face, it has been presented to me that I was 'taking [an] extra break without recording it.' In fact, I was dealing with irate or disgruntled customers to the company's benefit. Of course the assumption was that I was stealing from the company."

Telematics may seem intrusive, but it is nothing compared to the new brain surveillance system in China. Here, workers are given special caps, some with cameras attached, that monitor the brainwaves of the worker. According to an article in the *South China Morning Post*, "Concealed in regular safety helmets or uniform hats, these lightweight, wireless sensors constantly monitor the wearer's brainwaves and stream the data to computers that use artificial intelligence algorithms to detect emotional spikes such as depression, anxiety or rage." The technology is already in wide-scale use and has been deployed in the military, public transport, factories, and in state-owned companies. Proponents of the system argue that it has boosted efficiency and workers make fewer mistakes. Opponents say that even emotions must be limited for high productivity. It is turning human workers into machines.

We are being watched, indoors and out, at work and at home. There is no sphere in which we are free of surveillance. And while our fears tend to be directed to hackers spying through baby monitors, or peeping Toms peering through our windows, the biggest window into our private worlds stares right at us every day: the black pinhole of our webcams.

In 2014, Edward Snowden revealed that the United Kingdom's GCHQ (Government Communications Headquarters) had been tapping into British citizens' home webcams under a program called Optic Nerve.[6] For six months in 2008, over 1.8 million Yahoo! chat accounts were compromised as agents

6 The optic nerve, it's worth remembering, transmits visual information to the brain, and is also the location of the human blind spot in each eye. It allows us to see, but its location is also hidden from our sight.

siphoned millions of images through home laptop and desktop computer cameras. In this most private of spheres, ordinary, innocent citizens were the targets. The system gobbled up whatever was before it, snapping a photo once every five minutes. This was a test bed for facial recognition experiments.[7] People at home were unaware, of course, that they were being watched by the government, and as a result 11 percent of the images captured by officials contained nudity and were marked as explicit. Snowden's leaked documents revealed only what was swept up in half a year.

Together, Australia, Canada, New Zealand, the United Kingdom, and the United States form the Five Eyes intelligence alliance, with capabilities to surveil vast populations across the globe. We still have no way of knowing how much access they have to our private video and audio communications, but we do know that their systems grow more sophisticated, more expansive, and more intrusive every year.

The Eyes in Our Heads

TODAY, THERE ARE EYES in the sky, eyes that surround us on the ground, and even eyes prying into our minds. Social media is, on its surface, where we post content to connect with others in our lives. It's where we share images of our babies and pets, our food and vacations, as well as our likes and dislikes, our dreams and aspirations. But for the companies that collect our data, our profiles are really digital dossiers. They reveal our preferences, our sexual orientations, our religious outlooks, and our political

---

7 Data from public records has found a 92% false positive rate for facial recognition: "public records request shows that of the 2,470 alerts from the facial recognition system, 2,297 were false positives. in other words, nine out of 10 times, the system erroneously flagged someone as being suspicious or worthy of arrest."

alliances. Kept on file, they prove that what we say can be used against us, even if we haven't been arrested. This is why in 2018 the US State Department proposed a new form for all visa applicants to the country. In an effort to siphon more data for background checks, applicants must now submit a full list of their social media handles so that government officials can prescreen them on their various social media platforms. That our public posts are scanned by police and other authorities is not new. A 2013 survey by the International Association of Chiefs of Police found that 96 percent of police forces use social media in some capacity, with the most common use (86 percent) being for criminal investigations. But in a wide-sweeping dragnet there are always false positives. In the documentary *Terms and Conditions May Apply*, New York comedian Joe Lipari felt the effects of this first-hand when he paraphrased a line from the film *Fight Club* and posted it to Facebook.[8] Say the wrong thing and there can be serious consequences. Two hours later, a SWAT team arrived at his door, and he spent a year in court in order to prove he was not, in fact, a terrorist. As a U.S. Army vet, Lipari said, "I always thought we were on the side of the people. Now, I see the government looks at all of us as potential threats, no matter how meritorious our civilian lives or military careers."

But it's not just what we post in public that's monitored. It's what we look at privately as well: our searches, our likes, our posts, and our ad clicks are just a part of the digital trail we leave

8 He posted, "Joe Lipari might walk into an Apple store on Fifth Avenue with an Armalite AR-10 gas powered semi-automatic weapon and pump round after round into one of those smug, fruity little concierges." The original quote from *Fight Club* is: "And this button-down, Oxford-cloth psycho might just snap, and then stalk from office to office with an Armalite AR-10 carbine gas-powered semi-automatic weapon, pumping round after round into colleagues and co-workers."

behind. And while these cookie crumbs may be forgotten by us after an hour online, what we have said or thought is not lost at all; it is all saved as demographic and psychographic data on the servers of the sites and apps we've visited and used.

IBM has estimated that each day, the average person leaves behind a five-hundred-megabyte digital footprint. That was in 2012, ancient history compared to the data collected now. With people and objects increasingly connected to the internet through Fitbits, smartwatches, and connected homes, for example, IBM's estimate is likely a fraction of the digital footprint we leave behind today. That's because the vast majority of the world's information—some say up to 99.8 percent—was created in the last two years. According to one study, the digital universe will contain forty-four trillion gigabytes of data by 2020, a total output of at least 5,200 gigabytes for each person on Earth. As Stanford University professor Michal Kosinski puts it, if you imagine just *one day* of humanity's data printed out on paper, double-sided in twelve-point type, the stack of paper would extend to the sun and back four times. That is a mind-boggling amount of data.

And what value does this data have? According to *The Economist*, "The world's most valuable resource is no longer oil, but data." In the first quarter of 2017, Amazon, Apple, Facebook, Google, and Microsoft netted $25 billion in profit. Amazon alone accounted for half of all online dollars spent in America. Our data is valuable because that's how we are targeted as consumers. From the moment you turn on your computer or smartphone and begin to browse the internet, you are being invisibly followed. The new economic model is what Harvard Business School professor Shoshana Zuboff calls "surveillance capitalism." As she writes, "The game is selling access to the real-time flow of your daily life—your reality—in order to directly influence and modify your behavior for profit. This is the gateway to a new universe of monetization opportunities: restaurants who want to be your destination. Service vendors who want to fix your brake pads. Shops who will lure you like the fabled Sirens."

This is all done through personal data markets, the new primary business model for over a thousand companies. While we often hear of the business model of the FAANGs (the acronym given to the group that includes Facebook, Apple, Amazon, Netflix, and Google), Sarah Spiekermann, co-chair of the Institute of Electrical and Electronics Engineers' standard on ethics in IT design, reminds us that it's not "just Facebook and Google, Apple or Amazon that harvest and use our data. . . . Data management platforms (DMP) such as those operated by Acxiom and Oracle BlueKai possess thousands of personal attributes and socio-psychological profiles about hundreds of millions of users." From the moment we go online, a vast and sprawling apparatus swings into action, harvesting and sending our profiles to distant servers so that a tailored ad can be sent to us in milliseconds.

Unless you are in the advertising profession, you most likely have not heard of real-time bidding, or RTB, but it is how approximately 98 percent of all available ad space is sold on the internet. The automated bidding platform is similar to the Nasdaq, but instead of buying and selling stocks, it buys and sells you and me, or more specifically our data. For marketers, this data is digital gold.

This is how it works. Data management platforms house and aggregate what's known as first-, second-, and third-party data. In its simplest form, first-party data is data that comes from a website's own source—that is, its own visitors browsing and purchasing data and demographics. Second-party data is consumer data that is siphoned from a website's partner through partnership agreements. And third-party data comes from any number of aggregate outside sources that bundle and sell our data together.

Next the data is crunched and segmented into "audiences" to provide detailed markets: say a male sports fan, aged eighteen to twenty-five, who lives in the Toronto area and has made over $500 in online purchases within the last thirty days. That data is then exported and made available for real-time bidding. When the ad exchange is searching for a twenty-year-old male in the Toronto area whose regular searches include "hockey," this ID will pop up, allowing a sports e-tailer (a retailer selling goods on the internet) to bid on his impression so that their ad for hockey sticks shows up on websites he visits.

As for the specifics, what kicks into action is the demand side platform. As Pete Kluge, the group manager of product marketing for the Adobe Advertising Cloud, describes it, the "platform bids on each individual impression through RTB based on what is known about the user . . . so once that bid request comes from an ad exchange the demand side platform evaluates all the data that is known about that user and then determines the right price to bid for that individual user impression." This all takes place at lightning-fast speed and is almost imperceptible; from the moment you begin browsing to the time it takes for the ad to be served is approximately ten milliseconds.

Our data is not only hugely valuable for the surveillance economy, it is a veritable bonanza for political purposes as well. In March 2018, Christopher Wylie, a former director of research at Cambridge Analytica, came forward to reveal that the company had harvested the Facebook profiles of eighty-seven million people, whose data was used to sway both the United Kingdom's Brexit referendum and the 2016 US election.

All of this began as a simple offer on Mechanical Turk, an Amazon platform that allows users to make small amounts of money from participating in mundane tasks like web surveys, labelling objects in images, or watching internet videos. In this instance, the workers were asked to install a personality quiz app called "This Is Your Digital Life" for $1 to $2 per download. The requirement was simple: all you had to do was take the personality test while logged into your Facebook account. Over 270,000 people downloaded the quiz, approximately 32,000 of whom were US voters.

Users were told that the app was "a research tool used by psychologists," but it didn't just gather information on each test-taker who downloaded it. It also snaked through their friends' accounts and their entire digital community employing an algorithm that targeted hundreds of data points per person. The data was then used to create micro-targeted psychological profiles. As an article in *The Guardian* noted, it was the digital footprints that Cambridge Analytica was after: "The algorithm . . . trawls through the most apparently trivial, throwaway postings—the 'likes' users dole out as they browse the site—to gather sensitive personal information about sexual orientation, race, gender, even intelligence and childhood trauma."

Users were broken down into types—fearful, impulsive, or open, for instance—and political messages were created to target those specific attributes and begin to alter voter behaviour en masse. The goal of Cambridge Analytica was to persuade the undecided by using their psychology against them. Ultimately, they hijacked user data so that they could "not only read minds but change them." That's because, just as social media "likes" are used by advertisers to sell products to us, our profiles can also be used to learn how we can best be targeted. As Wylie suggests, while people may present different versions of themselves to their families, friends, or bosses, computers are neutral; they capture the digital trail of all of our personas. And because they have this complete picture, Wylie says, "computers are better at understanding who you are as a person than even your co-workers or your friends."

In the digital age, our lives have become open books. Users outside the United States have for some time had the ability to find out what Facebook knows about them. The result varies from person to person depending on how long they've been on Facebook and how active they are. Austrian law student Max Schrems tested this out by requesting all of his personal Facebook data. He received over 1,200 pages, a literal book of data, "including old chats, pokes, and material that had been deleted years before." The term "book" is a benign way of talking about our data; it makes it sound nice, kind of like a digital autobiography. Instead, it may be more helpful to call it what it really is, what each of us have on file is a record.

At least with social media companies, to some degree you opt in. Even if people do not know how their data is being used or what for, they have a sense that there is a trade-off, that they are giving something up to be able to use the service for free. Children, on the other hand, increasingly do not have that option. Since 2014, concerned parents in Colorado have been fighting what's known as the Golden Record, essentially a detailed data profile of their child that starts in preschool and follows them all the way through college. That record is a data pipeline that charts information and behaviour over the student's entire academic career: from test scores and attendance to family financial information, demographics, learning disabilities, mental health issues, and remedial courses and counselling or other interventions.[9] So what happens to all that data?

Education department officials say that its purpose is to "help guide parents, teachers, schools, districts and state

---

9 Every other year, students are given the "Healthy Kids Colorado Survey," which pries into personal details, asking, for example, "how old you were when you first had sexual intercourse, have you ever been molested, how many times you've driven a car when you'd used marijuana."

leaders, as we work together to improve student achievement so all children graduate ready for college and career." This sounds innocent enough were it not for the fine print that says that the data is also made available to "contracted vendors with signed privacy obligations for specified applications."

In a 2013 video, Dan Domagala, the chief information officer of the Colorado Department of Education stated that the longitudinal information can be shared in a "hub and spoke approach" and connected with other state agencies, including Human Services and the Department of Corrections. While the Department of Education insists that the data is aggregated and no identifying information on students is given, this is of no consolation to the parents. As one parent noted, all of this extensive collection of information on her own child is inaccessible to her, and "no parent has ever been able to access their child's Golden Record."

The Eyes on Our Bodies

THE POLICE ARRIVED at Sylvan Abbey Funeral Home unannounced. The previous month, in March 2018, officers had shot thirty-year-old Linus Phillip Jr. at a gas station. They had pulled him over for having tinted windows. The officers said they also found drugs in the car and that when Phillip tried to flee during the search, they shot him four times. Now the officers had come to the funeral home because they wanted access to his phone. They demanded that Phillip's corpse be pulled out of cold storage and used his finger to try to unlock his iPhone.

The attempt did not work, likely because the iPhone's fingerprint sensor reverts to a passcode after forty-eight hours. But the family was outraged by the brutal invasion of privacy. For Phillip's fiancée, not only had her partner been wrongfully murdered, he was now being "disrespected and violated" at the funeral home. The Florida police, however, had a more clinical approach. According to the law, they said, the dead did not have privacy rights. This is true. Using dead people's fingerprints in police work is fairly routine.[10]

To protect the public, security experts have been working on creating an "un-hackable" biometric system. For many, this is the holy grail, but to date every effort to create biometric security has been outstepped. That's because when the human body becomes the password, someone will find a way to crack it. In Malaysia, one gang was more literal than most when it came to their hacking method. When stealing a Mercedes-Benz that required unlocking using a fingerprint sensor to start the engine, they hacked off the owner's index finger with a machete so they could drive it away. There are of course subtler (and thankfully less violent) methods. At Michigan State University, experts have taken 2-D fingerprints, printed them on conductive paper so that they showed electrical conductivity, and unlocked phones. Every year, however, the biometric readers get more complex. The latest fingerprint readers use infrared scanners to detect vein patterns under the skin's surface and require active blood circulation to certify a person's identification.

In the Reality Bubble, the average person may not know much about how the system works, but the goal of the system is to know everything it possibly can about the average person. Biometrics is the final frontier of surveillance: the human body. It allows machines to read and identify human beings. Fingerprints are not just required of criminals or convicts anymore, because in a sense we are all suspects now. Fingerprints are used to unlock phones, provide access to airport lounges, and allow stock traders to use their mobile apps. Other parts of our bodies are being mapped and entered into databases as well. The NEXUS system uses iris scanners for trusted travellers to

---

10 Police can hire a company like Cellebrite to unlock the phone and bypass biometric readers as well. The cost is $1,500–$3,000 per phone.

clear customs quickly. MasterCard offers facial recognition so that people can "snap a selfie" and shop with their face. Banks use voice recognition to verify account holders. And now even our silence is under surveillance: you can be identified and tracked by the way you breathe. Scientists have found ways to "fingerprint" inhalations and exhalations, because our vocal passageways and lung capacities are individual to each of us. It is possible to use an algorithm to match a person to their "inter-vocalic breath sounds" over the phone, meaning we can be monitored and identified even if we don't speak.

The big question is, *why* are there eyes above us, around us, in our minds, and on our bodies? Why as a society are we being so deeply and intrusively surveilled?

IBM was the first company to use its tabulator machines to sort people into groups. In 1933, the company formed a twelve-year partnership with the Nazis, and its simple punch cards, called Hollerith cards, were used to organize and segregate concentration camp prisoners. As Edwin Black, author of *IBM and the Holocaust*, writes,

> The codes show IBM's numerical designation for various camps. Auschwitz was 001, Buchenwald was 002; Dachau was 003, and so on. Various prisoner types were reduced to IBM numbers, with 3 signifying homosexual, 9 for anti-social, and 12 for Gypsy. The IBM number 8 designated a Jew. Inmate death was also reduced to an IBM digit: 3 represented death by natural causes, 4 by execution, 5 by suicide, and code 6 designated "special treatment" in gas chambers. IBM engineers had to create Hollerith codes to differentiate between a Jew who had been worked to death and one who had been gassed.

Even this early system was subject to hacking. And the first ethical hacker was René Carmille, the comptroller general of the French Army, who headed up the French census before the Germans invaded. The Germans instructed Carmille to input census data into IBM machines and have it analyzed to produce a full list of Jews living in France. Carmille and his team had a different idea. They hacked the punch card machines so that data could not be entered for the column that specified religion. His sabotage worked until 1944, when the Nazis discovered the plot. Carmille was tortured and sent to the Dachau concentration camp, and he died shortly thereafter.

As digital law expert Heather Burns notes, this one small hack to the system had a lasting legacy: "In the Netherlands, 73% of Dutch Jews were found, deported, and executed. In France, that figure was 25%. It was that much lower because they couldn't find them. They couldn't find them because René Carmille and his team got political and hacked the data."

Data is how we segment and survey society. David Lyon, author of *Identifying Citizens*, makes the case clearly. Why are we turning people into data? Because "identification is the starting point of surveillance." It allows systems to sort people into groups that can by analyzed, classified, and, according to whatever data has been collected, rewarded or discriminated against.

The philosopher and economist David Hume famously remarked in 1741 that "nothing appears more surprising . . . than the easiness with which the many are governed by the few." And it *is* surprising. We take orders because our lives have already been ordered in ways that we rarely think about. Every morning, millions of people seemingly automatically get in their cars to commute to work, because time has become so ingrained in our minds that we do not question it; we simply obey it. In a similar vein, imagine how baffled First Nations people were when they first saw that American troops stop at the Medicine Line, the Canadian border at the forty-ninth parallel, while they could pass through. To their eyes, it was as though the troops had to stop because they were possessed by magic. But the effect was not so much magic as it was magical thinking. Today we all obey these imaginary lines because the importance of borders has been ingrained. As Yuval Noah Harari puts it in *Sapiens*, "People are willing to do such things when they trust the figments of their collective imagination." But these systems are no longer just imagined; like Pinocchio, the creations are becoming real.

If, in the past, the day-to-day march, the orders and controls, were accomplished through hegemony and the acquiescence of our beliefs, today our collective beliefs about how clocks and borders divide up the dimensions of the world have been formalized into a grid by which our lives are plotted. This is the infamous "system" that is everywhere but remains hidden to us, and unseen. Here, our measurements of time and space have entered the real world as data and our digital doppelgängers are placed under constant surveillance; we have become the "dots" on the grid, leaving a digital trail behind. In this virtual world, all of us are being followed. But if there's one element missing from the grid, it is our own flesh and blood. To truly integrate our bodies with the system, we too must become data. Biometrics provides that digital leash, allowing our real, physical bodies to be monitored and controlled in the real, physical world.

The Eyes with Their Own Mind

FOR THE POLICE, searching for Mr. Ao was like searching for a needle in the proverbial haystack. Over fifty thousand people were buzzing with excitement at Jacky Cheung's concert in the city of Nanchang in East China, and there, huddled in the crowd, Ao and his wife felt safe. But not long after the couple arrived, facial recognition cameras began scanning the pop concert arena. Moments later, Ao was spotted and arrested.

The newspapers reported that he was taken away and charged with committing unspecified "economic crimes."

Not only has China got a supercharged surveillance system, it is tied to a person's credit history as well. If, in the West, people obsess over social media "likes," in China people have an added obsession: their social credit scores. The system, proposed by China's State Council in 2014 to establish "a nationwide tracking system to rate the reputations of individuals, businesses, and even government officials," has been called the "gamifying [of] good behavior." As Mara Hvistendahl writes in *Wired* magazine, "The aim is for every Chinese citizen to be trailed by a file compiling data from public and private sources by 2020, and for those files to be searchable by fingerprints and other biometric characteristics."

Points are awarded for such meritorious behaviour as making early bill payments, stopping at crosswalks, conserving energy, giving to charity, and being friends with the right people. Having high-scoring friends in your social network is like a positive feedback loop, adding points to your own social credit score. And having a good score has benefits: a high-scorer can book a hotel room without paying a deposit, use free umbrellas, skip airport security lines, secure better interest rates at banks, receive energy bill discounts, gain access to apartment rentals, and even be rewarded with a boost to their profile on online dating sites. And increasingly, as facial recognition and technologies like "smile to pay" become integrated, a person's face will become tied to their social credit score as well as to their wallet.

These are the pluses. Points can be knocked off for bad behaviour, however. That can range from one's shopping habits to undesirable online speech. Points can also be docked for jaywalking, spreading fake news or anti-government propaganda, late payment of bills, loitering, illegal parking, playing too many video games, illegal home renovations, cheating on exams, or even just having a lower digital class of friends. And the penalties aren't frivolous. Low-scorers may have difficulty getting their children into good schools or getting mortgages or government jobs.

As Hvistendahl writes, with a low social credit score, you become, "effectively, a second-class citizen." She documents a low-scorer who "was banned from most forms of travel; he could only book the lowest classes of seat on the slowest trains. He could not buy certain consumer goods or stay at luxury hotels, and he was ineligible for large bank loans."

Low scores have put millions of Chinese citizens on travel blacklists. Already, four million have been blocked from high-speed train travel, and eleven million more have been blocked from buying airline tickets. And once people are on the blacklist, they often don't know how to get off. Even Zhang Yong, the deputy director of the National Development and Reform Commission, which regulates the social credit system, has stated that people "quite often" remain on the blacklist *after* their fines or debts have been repaid.

China's social credit system is still in its early phases. But the data pipe has been designed as a joint state-corporate system; for the most part, the government uses its high-tech infrastructure—such as facial recognition cameras to surveil and track citizens in public—while the bulk of the data compiled on citizens comes from private companies like Tencent and the Alibaba Group's Ant Financial that collect and track consumer behaviour, political allegiances, social networks, and payments (or nonpayments) on their apps.[11]

---

11 According to an article in the *The Sydney Morning Herald*, "The first phase of the national credit sharing information platform was being used by forty-four departments, across all provinces and sixty private enterprises, to disclose information and mete out 'joint punishment.'"

The irony is that historically in China the most powerful form of social control has been the desire to save face, or to keep up one's public image, because that reflects not only on the individual but on the respect given to and the dignity of the whole family. Today, that face is digitized data. And increasingly we are letting computers gauge our reputations based on our social media history. In the real world, our friends and families may in time forgive and forget, but there is no saving face with a computer, because a database will never forgive *or* forget.

With artificial intelligence, we hand over the reins to machines. As James Vincent writes in *The Verge* magazine, "We usually think of surveillance cameras as digital eyes, watching over us or watching out for us, depending on your view. But really, they're more like portholes: useful only when someone is looking through them. . . . Artificial intelligence is giving surveillance cameras digital brains to match their eyes, letting them analyze live video with no humans necessary."

Our bodies are just another topography to a computer, just another map. Modern facial recognition systems chart our faces like landforms. Our "landmarks" are defined and measured, from the ridges of our noses, to the depth and width of our eye sockets, to the forms of our ears. In this domain, it is not people but computers that recognize us. Systems, like those used on Apple's Face ID, project thirty thousand infrared dots onto a person's face, creating a unique topography that is instantly read and recognized by the phone's sensors. Computer technologies like DeepFace, used by surveillance systems, can create 3-D models of our heads from 2-D images such as photographs, so that we can be tracked even if we are moving our heads or a camera is pointed at us from a different angle. Additionally, surface texture analysis can be used to improve ID methods. Algorithms analyze a patch of skin for particular characteristics, like lines and pores, and create a unique "skin print," which can improve facial recognition accuracy by 20 to 25 percent. Even thermal cameras are being tested so that a person's profile and the shape of their skull are recognizable regardless of whether they are wearing a hat, scarf, glasses, makeup, or anything else to disguise themselves.

Often, when the topic of surveillance comes up, you'll hear people say, "So what? I'm not doing anything wrong. What do I have to fear?" To be clear, you don't have to *do* anything wrong; you can just *be* in the wrong category. Whether it's being Jewish, Christian, or Muslim, gay or trans, poor, sick, or disabled, or a jaywalker. And facial recognition systems can not only spot a person's age, race, and gender but also their sexuality. This "gaydar" has an accuracy rate of 91 percent, which could lead to serious abuses in countries where being gay can result in imprisonment or even the death penalty.

In the Chinese province of Xinjiang, which is home to China's Muslim Uighur population, over 18.8 million people were required to participate in the Physicals for All program in 2017. The biodata collected from residents included DNA samples, blood samples, fingerprints, and iris scans. All of the data is crunched together, along with surveillance materials, to rank people as "safe," "normal," or "unsafe." Opting out was not an option, and the desire to do so was regarded as a "thought problem" or "political disloyalty." Like the Golden Record system in Colorado schools, the involuntary participants are not given their results.

Today, in the main city of Ürümqi, one of the world's most advanced surveillance systems is in place. In an effort to root out violent separatists, as *The Wall Street Journal* notes, "Security checkpoints with identification scanners guard the train station and roads in and out of town. Facial scanners track comings and goings at hotels, shopping malls and banks. Police use hand-held devices to search smartphones for encrypted chat apps, politically charged videos and other suspect content. To fill up with gas, drivers must first swipe their ID cards and stare into a camera." For those who are blacklisted, an "X" shows up when their photo ID is scanned, and when that happens they are trapped. They can't travel anywhere.

The technology companies fetishizing and normalizing biometric technologies lure us into giving them our data with "cool features" like quick payments, VIP priorities, and silly avatar apps. But biometric data is not just surrendered voluntarily. In many countries it is being increasingly used without people's consent. In India, even newborn babies have their fingerprints scanned,[12] and in the United Kingdom four out of ten schools use biometric fingerprinting, creating a database of over 1.28 million pupils' fingerprints (31 percent of which were taken without the consultation of parents). It is also widespread in the United States, where hundreds of schools have begun using biometric fingerprinting for students to pay for their cafeteria meals.[13] When parents in the United States protested because of privacy concerns, some were told that unless they submitted their child to the procedure, "their children wouldn't be allowed to eat school lunches at all."

We are beginning to see the dark side of biometrics. When used to control access to basic life requirements, it has the power to penalize us in the most fundamental ways. You can be denied not only free movement but even the ability to eat. In India, the government has implemented the mandatory Aadhaar system, scanning the eyes, faces, and fingerprints of its 1.3 billion citizens.

---

12 In India, healthcare workers are using fingerprints as a primary form of ID for children without official documents.

13 The system can even be scaled up, "expanded, at no additional cost, to handle time and attendance, event admission, parking lot security and the tracking of students riding on school buses."

As Goel Vindu writes in *The Independent*, as part of the program, "The poor must scan their fingerprints at the ration shop to get their government allocations of rice. Retirees must do the same to get their pensions." Likewise, in Venezuela, over twenty-thousand fingerprint scanners were installed in supermarkets as part of the country's food rationing plan so that hoarding could be prevented and access to food, medicines, toilet paper, sanitary napkins, detergent, and other necessities could be either granted or limited.

Is It a Crime to See?

JEAN-JACQUES ROUSSEAU ONCE SAID that "man was born free, and everywhere he is in chains." The chains are not imaginary—they are very real—but to the naked eye they are invisible. And in this invisibility lies power. Our whole planet is surrounded by eyes. We are tracked by cameras from thirty-five thousand kilometres up in the sky and scanned and monitored right down to the very lines and pores in our skin. But here's the thing: we don't even notice we're being monitored. It's a huge blind spot. It's like we live in a high-tech panopticon, the circular prison building devised by philosopher Jeremy Bentham, where all the inmates could be observed in their cells without them ever knowing when someone was watching. The watchmen could see, but were not seen by those who were being watched.

Michel Foucault was aware of this when he stated that "disciplinary power is exercised through its invisibility . . . [and] at the same time it imposes on its subjects a compulsory visibility. It is this fact of being constantly seen . . . that assures the hold of the power that is exercised over them." Likewise, writing in *Guernica* magazine, John Berger stated that "the best way to understand the world is not as a metaphorical prison but a literal one." And that, without hyperbole, our current state of affairs is "nothing less. Across the planet we are living in a prison."

Throughout this book there has been one recurrent theme: in the twenty-first century, you'll find cameras everywhere except where our food comes from, where our energy comes from, and where our waste goes. These are the three blind spots of our human life-support system. And the system works to protect itself, which is why you'll find that it deliberately blinds us.

Ryan Shapiro, executive director of the transparency organization Property of the People, has been obtaining Freedom of Information Act (FOIA) documents for more than a decade. In one FOIA release, he found that one year after 9/11, despite all of the domestic terrorism priorities, the FBI was tracking vegans—specifically, at a Halloween party.

The document reads,

> SYNOPSIS: Vegan Halloween Party
> DETAILS: Philadelphia obtained an Internet posting of the Stop Huntingdon Animal Cruelty (SHAC) USA website. The posting advises of a Vegan Halloween Costume Party: . . . "On Saturday, October 19, starting at about 7:30 PM SHAC Philly will be having a Vegan Halloween costume party as a fundraiser at the Old Pine Community Centre (4th and Lombard). We will have a DJ, vegan food and a photo lab to take pictures of attendees and their dates in whatever holiday splendour they should choose to wear."

And why were FBI spies tracking an event that likely featured nothing more dangerous than quinoa and veggie patties? Because animal rights activists are seen as a direct threat to the food system. As we saw in Chapter Four, the conditions often aren't pretty. As Paul McCartney famously said, "If slaughterhouses had glass walls, everyone would be a vegetarian." But it isn't just walls that keep the reality hidden. It can be illegal to look in. Food companies who want to be shielded from public scrutiny have put forward ag-gag laws (short for agriculture gag) designed to prohibit undercover investigations on factory farms. First proposed in 2011, these laws would make it a crime to "produce a record which reproduces an image or sound"[14] inside an animal facility or even "possess or distribute" such a recording."

Distressing animal abuse videos and unsanitary farming conditions have led to meat recalls, which is bad for big business. But for doing the dirty work of exposing inhumane and sometimes horrific practices, poor food safety, and abuse of workers' rights, animal rights activists who go undercover to take videos and photographs and document abuses are considered "eco-terrorists" committing a felony, and if caught they can be fined or imprisoned.

The news often portrays animal rights campaigners as violent bomb-setters, so Will Potter, author of *Green Is the New Red*, decided to go to the source and research their history. Potter contacted the Foundation for Biomedical Research (FBR), a non-profit based in Washington, DC, that supports animal testing in medical and scientific research and is currently the only group in the world that keeps tabs on the crimes of eco-terrorists. If any group has the incentive to expose the crimes of animal rights activists, it's the FBR. But what Potter discovered was surprising: "The list of top eco-terrorism crimes from one of the top adversaries of these movements does not include a single injury or death." And while "thousands of violent criminal acts" have been documented in recent decades, he continues, a report investigating the subject "lists ninety-five crimes from 1984 to 2002, including multiple 'pie-ings.' A pie-ing is exactly what the name implies. Think Larry, Moe

14 Where ag-gag laws have been struck down, activists are being charged through other means. Rescues of sick or injured animals, for example, fall under federal theft with a charge of five years.

and Curly imbued with the revolutionary spirit of Situationist International."

It is illegal to look too deeply at where our food comes from, and the same is true of our energy and waste systems. Businesses and governments can spy on us, but we are forbidden from spying on them, and in some cases even forbidden from openly recording a public protest.

In October 2016, documentary filmmakers Deia Schlosberg and Lindsey Grayzel were arrested for turning on their cameras to film pipeline protests. Schlosberg faced three felony conspiracy charges and up to forty years in prison for recording events at TransCanada's Keystone Pipeline site in Pembina County, North Dakota, while Grayzel was subject to a strip search and jailed for filming a pipeline protest in Skagit County, Washington. After massive uproar and celebrity protests, the charges were dropped. Interviewed at the Portland EcoFilm Festival, Schlosberg told the audience, "Lindsey and I had never experienced anything like that: being arrested for doing our jobs, charged with felonies for doing our jobs." And as festival director Dawn Smallman put it, "If they could arrest Lindsey and Deia, they can arrest pretty much every filmmaker we show here at the EcoFilm Festival. This is a huge chilling thing if you work in media, if you work in film and if you're taking on big issues like climate change and the big corporations."

Compared to Vietnamese activist Hoang Duc Binh, the American filmmakers had it easy. On February 6, 2018, Binh was sentenced to fourteen years in prison for the crime of documenting fishermen who were protesting waste pollution. After a massive chemical spill by a multi-billion-dollar steel plant conglomerate devastated fishing communities and caused mass fish deaths along two hundred kilometres of coastline, the locals, whose livelihoods were dependent on the fish stocks, held a demonstration.

Binh's crime was live streaming the fishermen's protests on Facebook. According to the People's Court in Nghệ An Province, he was convicted of "abusing democratic freedoms to infringe on the interests of the state, organisation and people, and opposing officers on duty." During the live stream, Binh had told viewers that the fishermen were being beaten by authorities. The court ruled his statements were slander, while Binh denied wrongdoing. In essence, Binh's real crime was refusing to participate in official blindness.

In recent years, the silencing of environmental activists—those who document wrongdoings with respect to our food and energy production and disposal of our waste—has seen a steady increase. In 2017 alone, 197 environmental activists were murdered for exposing systemic abuses. According to the non-profit Global Witness, which documents crimes against activists, in 2016 up to four people were killed every week of the year.

Surveillance, then, is the means by which our modern life-support system is maintained. It ensures that we work efficiently and productively, that we are good consumers and shoppers, that we don't rock the boat or stray from the status quo. We are tracked on a grid that no longer requires that we "believe" in the hegemony of time and space. The new system is a physical manifestation that seeks to control and limit our behaviour through a physical apparatus.

It would be a mistake, however, to think there is any evil mastermind behind it all. There is no Big Brother. We are all watching over each other to ensure that we keep in line. We tell ourselves it keeps us safe, that surveillance protects the good people in society by tracking down the bad who engage in criminal behaviour. But ordinary people are being monitored and penalized too, for the most minor of "infractions." And modern surveillance is also used to track down and silence activists, the people attempting to turn the cameras back on the system to show us where it is going wrong.

That, perhaps, is what is most frightening. Our system of producing food and energy and disposing of waste is operating on a scale that is beyond alarming. We are prisoners to a system that if left unchecked will, it is no exaggeration to say, destroy most of life on Earth. And yet surveillance encourages us, indeed forces us, to keep on with business as usual, to turn a blind eye and look away.

10

THE EMPIRE WEARS NO CLOTHES

> Cursed is the first person who said, "This is mine."
>
> —CROATIAN PROVERB

The biggest bank holdup in history was invisible. There were no guns, no robbers in ski masks, and no trembling tellers forced to open the vault. That's because the money was not stacked in one physical place. Instead, it was stolen when it was on the move, flickering across continents in streams of ones and zeroes.

Employees at Bangladesh's central bank noticed something unusual on February 5, 2016. The printer that automatically printed out receipts of daily transactions had been completely idle. After some troubleshooting, the problem was finally resolved later in the day, and the printer began spitting out reams of messages from the Federal Reserve Bank of New York. They were questioning the transfer orders coming from Bangladesh, as they added up to a massive sum: a total of $1 billion.

Money in the modern economy moves at the speed of light, and unlike human beings it can easily cross most borders. The SWIFT (Society for Worldwide Interbank Financial Telecommunication) network is the method by which bank funds make their way around our planet. On an average day, about $5 trillion circulates through partner banks in the SWIFT network, which includes eleven thousand financial institutions in over two hundred countries and territories.

The thieves hadn't just hacked the bank with malware, however. They had also pulled off a sophisticated hack of time and space. It took from Thursday afternoon, New York time, until Tuesday morning, Bangladesh time, for bank officials to even figure out they had been robbed. That's because the hackers cleverly played the countries' time zones and geography against the bankers. As *The New York Times* reported,

> When the Fed had received a total of 70 fraudulent payment orders to four bank accounts in the Philippines and one in Sri Lanka, totaling $1 billion, Bangladesh Bank was closed for the weekend. On Sunday, when the bank reopened and discovered the error, it was unable to reach the Fed. [The director] sent a stop-payment order to the Philippines central bank, which was closed for the Chinese New Year. . . . Late on Monday afternoon, Dhaka time, as the Fed was opening for business, Bangladesh Bank asked officials in New York to block the money transfer to the Philippines but were told it was too late and that the money was with recipient banks.

In the end, the bank robbers did not get all the loot they'd hoped for. Instead, they managed to launder $81 million through casinos in the Philippines to offshore accounts. To date, the culprit has not been identified, but based on data trails the key suspect is North Korea. If it's true, one of the poorest countries in the world has taken to robbing banks.

But this brings up a key question: why is there such a massive division between rich countries and poor countries? According to Jason Hickel, author of *The Divide*, it has a lot to do with how money flows. In the eighteenth century, he observes, people in Asia actually had a higher standard of living than people did in Western Europe. And from Latin America to India to Africa, it was colonization that caused steep declines in standards of living and income. Raw materials were extracted from the land with cheap labour so that manufactured goods could be sold back to the colonies and to Western countries, while high import tariffs ensured there would be no significant competition to the colonial masters. In devaluing colonies' resources and the labour that produced it, colonialism created a system of "unequal exchange," an outflow of tremendous riches from poor countries that has played out over centuries and keeps billions poverty-ridden today. As Hickel writes, "In the past, colonial powers were able to dictate terms directly to their colonies. Today, while trade is technically 'free,' rich countries are able to get their way because they have much greater bargaining power. On top of this, trade agreements often prevent poor countries from protecting their workers in ways that rich countries do. And because multinational corporations now have the ability to scour the planet in search of the cheapest labour and goods, poor countries are forced to compete to drive costs down."

The reason it's more expensive to buy a shirt made locally than it is to have one shipped to you that was made on the far side of the planet is because of this unequal exchange. If the true costs were accounted for, the wealth drained from poor countries and transferred to rich countries has been estimated to total $4.9 trillion annually.

But there are other ways to rig the financial system. And concealing money is a well-known one. According to Global Financial Integrity, a non-profit that tracks the manner in which money is secretly moved, illicit financial flows (IFFs) were estimated to be as high as $3.5 trillion in 2014 alone. And what are IFFs? They are illegal transfers—like money laundering, trade misinvoicing, or the use of shell companies—to shift funds from country to country in ways that are "typically intended to be hidden and unobservable."

Money can do this now because it isn't real. Currency used to be based on something solid, like cattle, or clam shells, or tobacco, but today money is abstract, a network of symbols attached to our identities—numbers attached to our ID numbers—that fly through deep-sea cables and zip through the wireless air as a vapour. The vast majority of the world's money is ghostlike, invisible.

Perhaps the best-known trick for hiding money, though, is one that uses the veils of borders and geography to shelter funds. The rich benefit most as they can afford lawyers who are familiar with the ins and outs. I'm talking, of course, about tax havens. In 2016, according to Gabriel Zucman, author of *The Hidden Wealth of Nations*, approximately $8.6 trillion sat untaxed in offshore accounts. For perspective, that's almost three times the size of the entire global technology market, currently valued at approximately $3 trillion. Leaks of the Panama and Paradise Papers have revealed just how common this practice is. In Canada, the sixty biggest corporations were found to have over one thousand subsidiaries in offshore zones. As a result, while regular Canadians have to pay their taxes or face charges, Canadian corporations and financial elites have found legitimate ways of keeping $15 billion off the books every year.

That's the general rule. Because money is free to move, the rich don't need to keep it at home. As *Fortune* magazine notes, in 2017 Apple Inc. held $252 billion of company profits offshore so it could avoid paying US taxes. Amazon made over $3 billion in profits, but thanks to tax exemptions and credits it paid almost nothing in federal taxes. In the same year, according to a report by Oxfam, a staggering "82 percent of wealth created across the globe went to the top 1 percent," and that 1 percent has enough combined wealth to end extreme poverty in the world seven times over. The gap between rich and poor has become so vast that the richest forty-two people on the planet have as much money as the poorest half of the world's population. That's forty-two individuals with the wealth of 3.7 billion people.

Being poor is not just about losing the game. The poor are increasingly penalized and criminalized for their lack of wealth. At the simplest level, many banks charge a fine, or a "fee," if an account balance is too low. The Bank of America, for example, has proposed a $12 fee for monthly balances below $1,500, in essence charging people for not having enough money. Banks also mine our data and use algorithms to determine credit scores. Even shopping repeatedly in "stores associated with poor repayments" can have an impact on one's credit, leading to reduced spending limits and a higher cost for borrowing money. And while the poor can't just pack up and move to a better neighbourhood, for the homeless it's even more difficult. As the National Law Center on Homelessness & Poverty, in Washington, DC, reports, "Despite a lack of affordable housing and shelter space, many cities have chosen to criminally or civilly punish people living on the street for doing what any human being must do to survive. Cities continue to threaten, arrest, and ticket homeless persons for performing life-sustaining activities—such as sleeping or sitting down—in outdoor public places, despite a lack of any lawful indoor alternatives."

In recent years, bans on sitting or lying down in public spaces have increased by 52 percent, and "anti-homeless spikes"—similar to the spikes you might see to prevent birds roosting—have been erected on park benches and flat spaces with the goal of making resting or sleeping there impossible. Even faith-based organizations and good Samaritans face arrest and criminal liability for feeding hungry homeless people. In the United States, over fifty cities, including Atlanta, Los Angeles, Miami, Phoenix, and San Diego, have anti-camping or anti-food-sharing laws.

The yawning gap between rich and poor is a function of how much access each group has to the system of exchange human beings use to trade: money. On a macro level, the global economy depends on the constant flow and exchange of money, but most people, including 84 percent of British lawmakers, do not know where money comes from. What many people consider "real" money, the physical currency of banknotes and coins, accounts for approximately $5 trillion, only 16 percent of all the money circulating in the world. According to the *CIA World Factbook*, the global total of money, or what's known as "broad money," is somewhere in the range of $80 trillion. So where does all this other money come from? It does not, as our parents liked to remind us, grow on trees. But it does grow on computers.

In a white paper called "Money Creation in the Modern Economy," the Bank of England explains that money is created through debt. Specifically, "Whenever a bank makes a loan, it simultaneously creates a matching deposit in the borrower's bank account, thereby creating new money." The way the textbooks explain it, the bank continues, is a fallacy: "The reality of how money is created today differs from the description found in some economics textbooks: Rather than banks receiving deposits when households save and then lending them out, *bank lending creates deposits* [emphasis mine]." Debt is a requirement for our modern economic system to keep functioning, because debt creates wealth.

On another level, we all know the origin story of money. We know that money is not "real"; whether it's a piece of paper or a coin or a digital transfer, money at its most fundamental level is an IOU. It's a promise. Global debt, however, has hit record levels, and these debts are surpassed every year. Globally, the world is sitting on $247-trillion worth of these largely empty promises, with a debt rate that has grown a jaw-dropping 40 percent in the last ten years.

To stay afloat, the world's poorest countries have had to put a mortgage on their futures. Since 1980, countries in the Global South have been paying off the interest on their debts to a tune of $200 billion a year. In total, according to Jason Hickel, that amounts to $4.2 trillion in interest payments alone that has flowed from the pockets of poor countries to rich countries. Rich nations, it should be noted, are also saddled with enormous debts, but these are owed to banks and individual investors, not, broadly speaking, to foreign governments. And where loans are made to rich countries, they come in the form of government-issued bonds with very low rates of interest. As Annie Logue explains in *How We Get to Next* magazine, "Rates charged to these [rich] countries to borrow money cover the use of the money and expected inflation, but they don't consider any repayment risk."

The old saying is that money makes the world go round, but it is the invisible, vaporous nature of money and the rigged rules of the game that spirit wealth away into the bank accounts of the wealthy while the debt load increases on the poor. And so perhaps we should consider a different saying when it comes to how we sustain ourselves in the modern economy. As a corporate villain in the TV series *Mr. Robot* once said, "Give a man a gun and he can rob a bank. Give a man a bank and he can rob the world."

AT THE INTERSECTION of South Finley Street and Dearing Street in Athens, Georgia, a grand white oak stands proudly, its branches raised up in a flutter of verdant leaves. Beneath them, protected in the shade, is a stone plaque that reads,

> For and in consideration of the love
> I bear this tree and the great
> desire I have for its protection for all time,
> I convey entire possession
> of itself and all land within eight feet
> of the tree on all sides.
> —William H. Jackson

The oak is famously known to town residents as the Tree That Owns Itself, or, more accurately, the Son of the Tree That Owns Itself, as the original tree was damaged in an ice storm in 1907 and finally fell in 1942. Several hundred years old, it had been beloved by Colonel William Jackson, whose childhood was spent playing under the towering oak. To protect it, Jackson decided in the early 1800s to deed the tree, and the property surrounding it, to itself.

When the original tree fell, local residents sprouted saplings from one of its acorns and planted one in the same spot. Today, the next generation of this tree stands firm and "free" in its inherited soil and is now over fifteen metres tall. Legally, in the state of Georgia, non-human beings do not have rights, but there has never been a challenge to the oak's independence, as in the minds of the locals, this one tree, above and beyond all the others, has won the right to own itself.

The idea that a thing like a tree could have rights may seem absurd. We tend to believe that rights or legal privileges should exist only in the human domain, especially since a "right" is a human construct to begin with. But trees are living beings. As

Peter Wohlleben argues in his book *The Hidden Life of Trees*, these are social creatures that parent their young, communicate with one another, experience pain, have a capacity for memory, and have sex. They are not inanimate; they are alive, and they exist within a silent but dynamic community.

Just as humans use an underground network of tubes and wires called the internet to communicate with one another, the forest likewise uses a "wood wide web." Individual trees communicate via fungal networks that connect them to other trees at their roots. Suzanne Simard, professor of ecology at the University of British Colombia, has found that using these mycorrhizal networks, trees can communicate signals of distress, feed and nurture each other with carbon, nitrogen, phosphorus, and water, and pass on defence signals and chemicals to protect their community from potential threats. We cannot see or hear it happening—just as, looking at a system of wires, you cannot see messages being passed back and forth on the internet—but trees are in a very real sense communicating with one another.

That non-human life forms have their own forms of intelligence is something that scientists are just beginning to understand. It is fortunate, then, that around the world, rights are increasingly being granted to non-human life forms in a legal effort to protect them. Nature may not have a voice per se, but by giving the natural world a "right" its interests can at least be defended in court and it can have legal standing.

In a historic ruling on April 5, 2018, Colombia's supreme court did just that and changed the status of the country's portion of the Amazon basin so that it became an "entity subject of rights," essentially giving the ecosystem the same rights as a human being. After years of rampant destruction from illegal mining, logging, and agricultural expansion, including drug crops, the Amazon was being robbed and its resources sold off. From 2015 to 2016 alone, deforestation increased by 44 percent to 70,074 hectares, or about the size of New York City. By granting the Amazon rights, this rainforest will have legal rights so that it can be protected and defended.

Similarly, in March 2017, the Whanganui River in New Zealand was granted the legal right of "personhood." For the Māori, who have sought respect and justice for the river for over 150 years, it has never been a "thing"; it has always been an essence of life, a vital part of the community. The indigenous peoples of the area—the Whanganui River *iwi*, or tribe—have a traditional saying: "I am the river and the river is me." Which, as we saw in Chapter Two, is supported by science: we do have a physical connection to the world even though with the naked eye you cannot see it. In *The Rights of Nature*, David R. Boyd explains the Māori philosophy called *whanaungatanga*:

> *Whanaungatanga* is actually broader than kinship in the sense that it relates not only to relations between living humans, but also to an expansive web of relationships between people (living and dead), land, water, flora and fauna, and the spiritual world of *atua* (gods)—all bound together through *whakapapa* (genealogy). In other words, the Māori believe that all things in the universe, living and dead, animate and inanimate, are related, going back to Papatūānuku (the Earth), and Ranginui (the sky). Thus all the elements of nature are kin.

What New Zealand's Environment Court recognized is this basic but powerful connection: people are made of the water; they drink water; and ultimately "to pollute the water is to pollute the people." Unlike Colombia's Amazon, however, the Whanganui was granted a dedicated committee of human representatives to defend its rights. Furthermore, if industry should wish to divert the water's natural flow, that could be deemed a violation and its guardians could petition the court to defend it.

The idea that forests, mountains, soil, rivers, and oceans are not merely human property has not gone unchallenged. While over three dozen US communities currently have rights-of-nature ordinances to protect their local ecosystems, big business has also brought in its own top-gun lawyers to defend *its* land rights as well.

In Pennsylvania, an ongoing David and Goliath battle between the Grant Township community and Pennsylvania General Energy (PGE) has been raging for over six years. The issue at stake is the Little Mahoning Watershed, home to a range of fish, freshwater mussels, aquatic insects, the eastern hellbender salamander, and, vitally for the locals, the key source of their drinking water. In 2014, however, as part of the state's fracking boom, PGE received permits to dispose of its wastewater in deep injection wells. For the community, whose drinking water comes directly from private wells, a daily underground injection of over 150,000 litres of toxic and radioactive fracking water represents a risk to their water supply they are unwilling to subject themselves to—especially considering earthquakes have been a well-documented result of the fracking process.

In court, PGE's lawyers argued that the idea that the Little Mahoning Watershed should have rights through the community ordinance was "absurd" and a "circus act," claiming that "a watershed lacks consciousness, intelligence, cognition, communicability, or agency. The Watershed cannot decide to intervene, cannot accept representation or engage with counsel as a client, and cannot appear in court or testify." On January 5, 2018, federal judge Susan Paradise Baxter ruled in favour of the corporation, calling the attempts by the Community Environmental Legal Defense Fund "unreasonable" and "implausible," and saying that such an approach "creates enormous expense to parties and taxes limited judicial resources." The executive director of the fund and the team's legal attorney were fined $52,000, to be paid back to PGE. Further disciplinary measures were also forwarded to the disciplinary board of the Pennsylvania Supreme Court for consideration, including the suspension of attorney licences and even disbarment.

But is fighting for the rights of nature really a mockery of the courts? The irony of the situation is that corporations are just as much of an artificial construct, if not more so, than a living ecosystem. As David Boyd points out, "Many of the same arguments [of PGE's lawyers] used to attack the watershed's standing are equally applicable to their own client. A corporation is a legal fiction, lacking consciousness, intelligence, and cognition. It is incapable of doing the things the corporate lawyers suggest an ecosystem ought to be able to do, such as testify in court. It is remarkable that PGE's lawyers could describe watersheds as 'artificial constructs' while simultaneously believing that corporations are real persons to whom rights naturally belong."

But legally, corporations are "persons." They have almost all the same rights as people do. In the United States, that includes equal protection, religious liberty, freedom of speech, freedom of the press, freedom from unreasonable search and seizure, the right to trial by jury, the right against double jeopardy, the right to counsel, and due process, to name a few. We've believed in this idea for so long, we barely question it. As Adam Winkler writes in his book *We the Corporations*, corporations gained their first rights in 1809, half a century before legal rights were championed for African Americans or women. But here's the rub: while corporations do have rights, without a physical body they do not suffer the same penalties as humans. A corporation can do wrong, but a corporation cannot go to prison.

At the same time, our closest biological cousins, chimpanzees, which share 98.8 percent of our DNA, have had their right to personhood denied by US courts multiple times. So why—if its okay for a corporation to be a person—is it not okay for a living animal or ecosystem to attain those rights? The answer is that we have relegated animals to the status of property, somewhat above the status of inanimate things, but not much.

Still, you can love your property. There are certainly dairy cow farmers, for instance, who show deep care and affection for their animals; not every farmer subjects their animals to factory-farm cruelty. But this does not address the relationship between cow and farmer, because the cow is still owned. It's property. The cow is not free.

And property, whether it's an object, a cow, or a slave, does not have right of movement without the owner's consent. "It" cannot change its conditions even if it's unhappy, because it has no rights. The key point here is that rights are incompatible with ownership when it comes to living things. After all, if rivers and chimpanzees have rights, what's next? Will our bacon and eggs demand freedom? Our lumber and paper? Our leather shoes and our wool sweaters? All of this life, or extinguished life, is defined as our property to do with as we please. To begin to question that fundamental authority of our ownership of life would be to upend our whole system of thinking. That's because the core tenet of our entire economic system can be eviscerated by asking one simple question, which is: What does it even mean to "own" something anyway?

JUST BECAUSE YOU HAVE paid money for something does not necessarily mean you own it. Martha Fuqua learned that lesson first-hand when she bought a box of assorted goods for $7 at a West Virginia flea market. Inside the box was a plastic cow, a brown leather doll, and a napkin-sized oil painting by one of the greatest artists that ever lived. Fuqua did not know it at the time, but it was a Renoir, and on having it appraised she learned its value was upwards of $100,000. When word spread about the discovery of the small masterpiece, however, another potential owner soon staked a claim: the Baltimore Museum of Art asserted that *Landscape on the Banks of the Seine* was their property and had been stolen from the museum in 1951. After hearing of the museum's claim, the Fireman's Fund Insurance Company also got involved. They had paid the museum a $2,500 settlement on the theft and so had a strong case for legal ownership themselves. In this instance, the object changed hands multiple times but had three different "owners." So, who owned the painting?

A judge ruled that the rightful owner of the Renoir was the museum. The case required a judge, as many property-related disputes do, because the question of who owns what is rarely simple. In some countries, property litigation accounts for as much as 66 percent of annual court cases. The disputes range widely, from who owns a dead man's emails to who owns human genes to who owns the family dog. The question of who owns what is critically important to us, because not only do the things we own define our social status, they define us internally as well.

The psychologist William James was the first to posit that we have a "blind impulse" to form attachments to things, that our objects and property become in a sense part of our material identities. That is, the things I call "mine" extend beyond my physical body outwards to my clothes, my family, my house, my garden, and my car. Each of us exists as the epicentre of our stuff, and while many of these things are inanimate, they affect our emotions. As James wrote in 1890 in *The Principles of Psychology*, "A man's Self is the sum total of all that he CAN call his, not only his body and his psychic powers, but his clothes and his house, his wife and children, his ancestors and friends, his reputation and works, his lands and horses, and yacht and bank account. All these things give him the same emotions. If they wax and prosper, he feels triumphant; if they dwindle and die away, he feels cast down, not necessarily in the same degree for each thing, but in much the same way for all."

Flash forward to the twenty-first century, and marketers and retailers are well aware that we see products and things as extensions of our physical selves. Studies have shown that merely touching an object is enough to create feelings of ownership, which is why people are encouraged to use samplers, to try on clothes, or take out cars for test drives.

The process takes place very quickly. When you walk into a store, you are fully aware that none of the objects there belong to you, but once you get to the cash register and pay for an item, the object instantly changes, not physically but in your mind. Now, it belongs to you. As Gail Wynand, the newspaper tycoon in Ayn Rand's novel *The Fountainhead*, put it, "I am the most offensively possessive man on earth. I do something to things. Let me pick up an ashtray from a dime-store counter, pay for it and put it in my pocket—and it becomes a special kind of ashtray, unlike any on earth, because it's mine."

Psychologists even have a name for this phenomenon. It's called the "instant endowment effect," and it marks the sudden attachment to an object we develop once we own it. The ownership effect is even visible in the brain. Using fMRI scans, scientists have found that when a person thinks of an object they own versus one owned by someone else, their medial prefrontal cortex lights up. This is the same area of the brain associated with "self-referential processing," and it activates when we hear our name, recall autobiographic memories, or remember our personal preferences.

Humans are certainly not alone in our drive to protect objects and defend territory; other animals are known to have a rudimentary understanding of exchange and possession. Bowerbirds, for instance, collect colourful trinkets to display by their nests, coconut octopuses will fight each other over coconut shelters, and high-ranking baboons respect the rights of possession of other troop members and will not take an object away from another baboon, even a low-ranking one, if he or she was in possession of the object first. So it seems there is some genetic basis for possession in the animal kingdom, but absolutely nothing that compares with the acquisitiveness of humans. While other animals have territories and shelters, we are the only species with such a vast number and variety of possessions. Most animals travel light; most humans are bogged down by their "stuff."

Our stuff may, however, be one of the keys to why our species is so powerful. As naked apes roaming the open savannah, we relied on our big brains for survival, and three hundred thousand years ago, some of the first objects we created were weapons such as arrows and spears. These became our first possessions. Hunters would have cherished and used their best weapons again and again, and from this perspective it makes sense that our early possessions were vital for survival.

As our species became increasingly sedentary, the drive to accumulate stuff went up. Archaeologist Gary Feinman argues that storing surplus was a way to minimize risk, because as "people settled down, they became more susceptible to environmental disaster." With families each stocking up on supplies, people's relationships were reinforced with trade, and the "exchange of non-necessary goods" became a way of strengthening ties with one's neighbours.

Today, ownership is considered a human universal. It is found in all cultures, though there is significant cultural variation. So while it could be said that ownership has evolutionary roots, much about ownership, particularly its rules, is learned rather than innate.

We begin to learn those rules early. As everyone knows, the concept of "mine" becomes very important to toddlers. Developmental psychologists have found that children as young as eighteen months can distinguish the difference between something they are in possession of versus something that belongs to them; and by two years they can reason that ownership belongs to the person who acquired something first. This notion of "first possession," as we shall see, is one of the primary ways in which adults define legal ownership.

To start, however, we should remember that ownership does not exist in a vacuum. It ceases to exist if there is no other person in the picture. From a psychological standpoint, "objects are claimed to distinguish them from something belonging to the other . . . Without the presence of the other, the need to label objects as 'belonging to me' or 'not belonging to me' disappears." Or to put it another way: if you were alone on a desert island, everything would belong to you and nothing would belong to you; it simply wouldn't matter. Until, that is, someone else showed up. Once there is another person in the picture, we can begin to assert our rights with respect to what we own. And children by the age of two or three can be seen defending the "rights" to their property. In a study where toddlers watched puppets throw objects into a garbage bin, for example, the children were fine if a puppet threw away its own property, but they vehemently protested when it threw away an object that belonged to them. Recognition of others' right to property begins around this age as well. Two-year-olds would watch without protesting if a puppet threw away an item owned by a third party, but by the age of three they would object at this violation of someone else's ownership rights.

So how do children decide what is theirs and what is not? First possession is the simplest rule, but it depends on how that possession was acquired. In an equal setting, with an object that belongs to no one, it's the kindergarten rule of "finders, keepers." As Shaylene Nancekivell, a cognitive psychologist at the University of Michigan, notes, this is why "collecting seashells lying on a public beach is acceptable, [while] helping yourself to seashells sold at a beachside stand is not." The vendor found them first and has the right to sell them. But in a world where most items are bought in store transactions, we seldom move beyond autopilot to think about who owns what. If one begins to consider the question, however, as cognitive development psychologists have, it becomes clear that "ownership is not a 'natural' property of objects, but is determined by human intentions" and that "facts about who owns what may be altered by appropriate decisions."

In the journal *Cognition*, Max Palamar and his colleagues explored this idea of ownership and intention by using the thought experiment of a feather sitting atop a cactus. In one scenario, someone called Mike wants the feather, so he uses a stick to dislodge it and it flutters down. Studies on this kind of ownership reasoning have found that most people will deem Mike the rightful owner of the feather. But if Mike bumps the cactus by accident and the feather falls just as Dave happens to walk by, and Dave picks it up, well then the rules change. Dave becomes the rightful owner of the feather because Mike had less direct responsibility for getting it.

History is another factor in deciding who owns what. As stated earlier, by the age of two or three most children already have a basic understanding of the ownership rule. To test this, toddlers were given identical toy cars to play with, and after playing alongside each other for a while, they had to tell the researcher which toy belonged to them. Two-year-olds had difficulty telling ownership since the cars all looked the same, but the three-year-olds were able to track their own toys by intently following the history of exchanges between their car and the cars of other children.

But history and primary possession alone are not the only ways we come to reason who owns what. If an object has an identifiable creator, that also affects our ideas of ownership. First possession, for instance, can be trumped by creative "investment" in an object. In a study looking at this behaviour, children were given Play-Doh and told that it was theirs to keep after the experiment was over. During the experiment, however, they were asked to swap their clay with the experimenter, and each would use the other's clay to shape an object. For the children, once the object was made, say a clay dinosaur, this new condition trumped first possession. Even though the clay was borrowed, the creative labour invested by the borrower made them the rightful owner of the object, and a transfer of ownership took place because of it.

An equal swap of free modelling clay is one thing, but ownership rights are trickier if the raw material is considered valuable, like gold. Looking at borrowed materials, researchers found there were significant differences between children and adults. Given borrowed wood to make a statue, most three- and four-year-olds believed they should get to keep the statue, whereas most adults did not. Furthermore, the greater the value of the loaned raw materials, the less impact creative labour had on ownership. When one of two materials was loaned to participants, either paper or gold, even if creative labour was involved, the finished product was thought to belong to the owner of the material if it was made of gold, regardless of the amount of time the other person spent creating the finished product.

As these cases illustrate, ownership is a slippery beast. There are no hard and fast rules, because we make and change them depending on the circumstance. To offer one final example, first possession can also be trumped by another factor: current use.

Ori Friedman and his colleagues wanted to find out if extenuating factors could change people's minds about an object's use. In a series of studies, adults and three- to seven-year-old children were told different stories about property disputes and asked who deserved the rights over the object. In one of the scenarios, a boy is using a crayon to make a card for his mother, but the crayon belongs to a girl, and now she wants it back. For children, the answer overwhelmingly was to defer to the rule of first possession and return the crayon to the girl. Adults prioritized current use of the object and believed the boy should keep the crayon to finish his project. When the crayon switched ownership and was said to belong to the teacher, a neutral third party, adults and children both said the boy using the crayon could keep it.

It's critical to investigate and untangle our ideas about ownership, because in the real world our property disputes extend well beyond the borrowing of crayons in school. On a geopolitical scale, exactly these arguments crop up over land rights, state boundaries, and historical ownership. And in high conflict regions like Israel and Palestine, or with respect to unceded Indigenous territory in Canada or China's historical claim in the South China Sea, these questions arise again and again. Do you own the land because you got there first? Because you are currently using it? Or because you added value to it and "improved" it?

The argument of "improvement" is the foundation for modern property rights and was put forward by the seventeenth-century philosopher John Locke. In 1690, he wrote in the *Second Treatise of Civil Government*, "As much land as a man tills, plants, improves, cultivates, and can use the product of, so much is his property." His logic is based on the idea that because we own our bodies, by extension we own what our bodies labour upon: "Though the Earth, and all inferior Creatures be common to all Men, yet every Man has a Property in his own Person. This no Body has any Right to but himself. The Labour of his Body, and the Work of his Hands, we may say are properly his. Whatsoever then he removes out of the State that Nature hath provided, and left it in, he hath mixed his Labour with, and joined to it something that is his own, and thereby makes it his Property."

It has been said that this passage has influenced the trajectory of Western civilization more than anything ever written. It came to define how people could own the land. According to Locke, the intermingling of labour and nature is where the magic happens. And thus, as he suggests, a person who picks an apple from a tree owns the apple. Today, however, while Locke's ideas serve as the philosophical basis for modern property law, we are a far cry from this simplistic notion. Modern fruit pickers working on farms, for instance, do not own the apples they pick; they belong to the farm, or to the corporation that owns the farm. At the supermarket, the apple belongs to the shopkeeper until we pay for it and then it belongs to us. Labour adds market value at every step along the way, also adding layers of complexity to the question of ownership. What goes unquestioned is the assumption that *anyone* can own it.

That was not always the case. Writing in the 1700s, William Blackstone, famous for his commentaries on English common law, challenged our ideas of ownership. That one person could claim the right over an object "to the total exclusion of any other individual in the universe" was not anything inherently natural at all, he argued. If we were to look at an object's history, we would begin to question the idea of that authority. "There is no foundation in nature or in natural law," he wrote, "why a set of words upon parchment should convey the dominion of land; why the son should have a right to exclude his fellow creatures from a determinate spot of ground, because his father had done so, before him; or why the occupier of a particular field or of a jewel, when lying on his death-bed and no longer able to maintain possession, should be entitled to tell the rest of the world which of them should enjoy it after him."

And yet, at heart, each and every one of us knows this already: that when we die we can't take our things with us. That's because our things are not a physical extension of our bodies; they are only an extension of our minds.

AT THE KŌFUKU-JI TEMPLE in the Chiba prefecture of Japan, a bizarre ritual has been taking place since 2015. If you were to visit on the right day, you would witness Buddhist monks holding a funeral for Sony robot dogs. It sounds like a PR stunt, but the ceremony is real. Real incense wafts over the dearly departed, a real priest chants traditional sutras, and real tears fall as the robots' owners say their final goodbyes.

The Sony Aibo (short for artificial intelligence robot) was designed to "get to know" its human companion. Trained to bark, do tricks, and respond to voice commands, it shaped its behaviour to meet its owners' preferences. As a result, some people grew very attached to their robotic pets, even coming to see them as family members. But in 2006, after Sony discontinued the product line, owners were left on their own with their Aibos, and problems mounted when older parts began malfunctioning. Seeing the need for repairs, a former Sony employee set up a "vet hospital" for robots that had begun to break down. And when a dog reaches the stage when it is beyond repair, there is one other avenue for it, as an "organ donor." Terminal robots have their functioning parts removed so that they can be donated to the robots that are still considered "living." The funeral, then, has become an important part of honouring the dead robots before they go to robot heaven after disassembly.

The ceremony is not just for the old hardware. As Bungen Oi, the chief priest of the service, has stated, "All things have a bit of soul." This belief in animism, that objects have a soul, is prevalent in Japanese custom and is also found in Shintoism, the country's primary religion. Similarly, in the Kantō region, on February 8 of every year, another funeral of sorts takes place as women gather together in colourful kimonos to celebrate the Hari-Kuyō festival. In Japanese, *hari* means "needle," and *kuyo* is "memorial." In essence, the ceremony brings seamstresses and kimono makers together to bury their old needles in tofu or jelly cakes, a final soft resting place for their lifetime of hard work.

These rituals are part of the larger belief that rivers, rocks, trees, places, and animals all possess a sacred essence, and that, similarly, everyday objects can have a spirit too. In fact, according to Shinto practitioners, after one hundred years, objects are said to acquire a soul. Household items like teakettles, dolls, and knives can come to possess what's known as a *tsukumogami*, or an "artifact spirit." Because of this, objects, whether they are toys, weapons, or tools, must be repaired and cared for so as not to offend the spirits that dwell inside.

To people in the West, the idea that things have spirits may sound absurd, but the same idea is common in the West as well. There are people who name and talk to their cars as though they were rational beings, just as others get angry at their computers and photocopiers. A writer may have a "special" pen, and a baseball player a "lucky" bat. People also worship religious icons, with some claiming that statues can even weep or bleed. And, of course, there are the cherished objects, like wedding rings or other personal items of significance. Indeed, a whole business has been built around psychics and mediums based on the belief that a part of the human spirit can be accessed by holding an object onto which something of the person who owned it has "rubbed off."

But what's really odd is that while we come to love some objects dearly and treat them as sacred, we toss away most of our things without a second thought. Imagine for a moment a database of everything you own right now. Include everything:

your house, car, clothes, shoes, bags, books, household appliances, jewellery, furniture, electronics, light bulbs, toiletries, trinkets, the items in your fridge, your music collection, right down to your last stick of gum. Now, try to imagine a second list that includes everything you have ever owned *and* everything you've ever thrown away. It's impossible to do, because the average person owns millions of things over a lifetime. We just don't get attached to all of them. "My chocolate bar" will be consumed and the empty wrapper tossed out, just as "my pen" will soon become unusable and unceremoniously thrown away.

The things we keep are things we can use, or things with sentimental value, like memorabilia, gifts, and heirlooms, which come with memories attached to them. We treasure these objects because they can transport us to a particular time and a place. It's why people have garages filled with stuff that they love but never use. This kind of mass storage is considered normal in a material culture.

Society does, however, have a name for people who don't part easily with disposable objects. They are known as hoarders. But hoarding is not, as some might think, a disorder stemming from materialistic values. Hoarders develop strong emotional attachments to almost all objects that belong to them. Their difficulty stems from impaired decision-making, as even useless objects are seen as an extension of self. As a result, hoarders do not know what to keep and what to throw out. In a sense, this is an acute version of a problem we all have to one degree or another: the idea that we are our things. That said, normal populations could be considered just as deviant if one considers the insane volume of stuff we throw away every day.

Historically, not all cultures have shown the same regard (or disregard) for stuff. And while hoarding is considered to be universal, with documented cases in societies around the world, on a cultural, rather than individual, level approaches to possession and materialism can differ dramatically.

Christopher Columbus was clearly taken aback by the people of Hispaniola and their approach to ownership. In 1493, in a letter back to Spain detailing his first voyage, he wrote, "They are so artless and so free with all they possess, that no one would believe it without having seen it. Of anything they have, if you ask them for it, they never say no; rather they invite the person to share it, and show as much love as if they were giving their hearts; and whether the thing be of value or of small price, at once they are content with whatever little thing of whatever kind may be given to them."

Likewise, Captain James Cook was awed by the behaviour of the natives of New South Wales in Australia, who did not covet material goods and were content that nature provided them with their basic needs. They had no need for "superfluities."

Modern capitalist societies see the role of goods completely differently. In our era, excess is a necessity. That's because the economy relies on growth, and growth relies on producing, consuming, and discarding ever more stuff. This is what Hannah Arendt called in 1958 the "waste economy." In *The Human Condition*, she writes, "Things must be almost as quickly devoured and discarded as they appeared in the world." As a consequence, the cycle of attachment is short. Rare indeed are the objects that should reach the hundred-year-old age of the *tsukumogami* spirits, since modern products are replaced or abandoned in a much shorter time.

It may be that we feel differently about our goods now as well. That's because most things we own aren't handmade. As a result, according to a study in *The Journal of Marketing*, they are missing the key ingredient of "love." We treasure handmade things, knowing that personal time and effort have been invested in their creation. It's why we love hand-knit sweaters made by our grandmothers, or simple crafts and drawings made by our children. Likewise, objects in the commercial sphere that are handmade are thought to "contain and transmit the artisan's 'essence.' . . . The customer then perceives the handmade product itself to be literally imbued with love."

Nowadays, the vast majority of our goods are made by robots and machines. Each plate, or sweater, or cell phone is a clone of the next. It's soulless, making it easier for us to part with it. On another level, as anyone who has seen the TV show *How It's Made* is aware, the scale and speed at which consumer goods are produced is dizzying. But this model of production also forces us into a perpetual loop. Machines don't get tired. They never complain about overtime. They are fast, efficient, and precise, and outperform any human when it comes to their endless ability to produce.

That leaves us with the corollary: to match the needs of hyper-intense productivity, we now have one key role, to consume.

THE CHAOTIC SCENE at the Wal-Mart in Porter Ranch, California, was captured on cell phone video. At 10:10 P.M., crowds waiting in line began screaming and yelling, wiping madly at their burning eyes while trying to flee. The crowds had gathered to purchase the latest Xbox 360. But as aggressive pushing and shoving began, an "unhappy customer," retaliating against her fellow customers, unleashed a canister of pepper spray.

If an alien were to observe our current state of affairs, they might note that the human species had gone crazy in its desire to own things. Over the years, Black Friday sales have become notorious for this type of behaviour, as shopping riots break out and people trample over each other and get into verbal and physical fights while competing to buy the latest electronics and home appliances. The website blackfridaydeathcount.com keeps track of these shopping fatalities. After 2018's Black Friday was over, the total stood at 12 deaths and 117 injuries.

Black Friday fisticuffs are in fact beginning to die down, as the buying frenzy is leaving bricks-and-mortar stores and shifting online. The growth here has been staggering. In 2013, Amazon alone sold 26.5 million items on Cyber Monday, or about 426 items per second. That number has now been eclipsed in China by a sales event called Singles' Day. What started as a kind of anti–Valentine's Day in 1993 by a group of Nanjing University students was co-opted by the e-commerce giant Alibaba in 2009 and transformed into a consumer marketing blitz. In 2017, Alibaba's Singles' Day sales hit $5 billion in the first fifteen minutes and reached over $25 billion in the course of the day, working out to 256,000 purchases per second.

This consumer boom, while healthy for the economy, is a catastrophic bust in terms of its physical consequences. Greenpeace Asia calculated that aside from the product waste, manufacturing, packaging, and shipping, the CO_2 from the 2016 Singles' Day shopping bonanza *in clothing alone* would require 2.58 million trees to absorb the emissions.

This out-of-control consumerism exacts a human toll as well. We are, quite literally, shopping ourselves to death. A recent study by Steven Davis and his colleagues at University of California, Irvine, found that 760,000 air pollution deaths annually are directly linked to consumer goods production.

From the outside, of course, this situation looks absurd. So it's worth asking ourselves why we do this. And the simple answer is we believe that having things makes us happy. But the happiness we get from material goods is only ever temporary. Planned obsolescence and the need to upgrade, stay fashionable, and maintain social status have us trapped like hamsters on a treadmill.

This concept of a "hedonic treadmill" was first put forward by psychologists Philip Brickman and Donald Campbell to refer to the process by which humans experience short-term mood shifts from external events but then quickly return to a set point, or baseline, for happiness. That's why we experience the initial thrill when buying a new product. As Derren Brown writes in his book *Happy*,

> I desire at the time of writing, the 6th incarnation of the Apple Macintosh internet-enabled smart telephone but I know it won't really make me any happier. After a short while, roughly equivalent to the time it takes me to explore its new features and get used to its new shape and weight, I will feel exactly the same about it as I do about my current one. Clearly, Apple know this and keep developing new models at such a pace that will make my non-ownership of the newest and best painfully obvious to me, adding a negative reinforcement to the process. There is the pleasure of the new model and the displeasure of knowing mine does not have certain features being enjoyed by everyone else. How pathetic.

Happiness is not the same thing as self-esteem. Researchers have found that people can have high self-esteem and be unhappy, or be happy and have low self-esteem. Perhaps unsurprisingly, studies have found that social media has a negative effect on self-esteem. As people compare and gauge their material status with others online, they begin to feel dissatisfied with where they are in the social hierarchy. Exposure to Instagram accounts where users flaunt their luxury goods, for instance, frequently creates a negative social comparison and lowers the self-esteem of the person browsing.

According to Tim Kasser, who has spent thirty years studying the psychology of materialism, "What research has shown in literally dozens of studies is that the more that people prioritize materialistic values, the less happy they are, the less satisfied they are with their lives, the less vital and energetic they feel, the less likely they are to experience pleasant emotions like happiness and contentment and joy, the more depressed they are, the more anxious they are, the more they experience unpleasant emotions like fear and anger and sadness, [and] the more likely they are to engage in the use of substances like cigarettes and alcohol."

We are crushing our own spirits in the quest to own better things, to own more things, things that ironically we soon won't want and will throw away. The worst part is, this dependence on acquiring objects tends to worsen when times are tough, because when we feel insecure, having something solid to cling to becomes a coping mechanism. Our possessions give us some semblance of control over the world. They give us power. We, the weak species of naked ape, did not rely on brute strength but rather on our brains to dominate. And we did it with our stuff. We became the masters of things. Things that made us stronger, faster, more powerful, better defended, more efficient, and more dangerous.

In the modern world, this power translates into action. We have the power to communicate with each other at the speed of light, to fly at the speed of sound. Individually, our things give us the freedom to shuttle ourselves independently across vast distances. Our things save us time. Machines wash our clothes and do our dishes. With the press of a button on a food processor you can chop food in an instant, where doing it by hand might take half an hour. Our things also save us from the grind of manual labour; instead, fleets of robots work overtime to assemble our mass-produced goods.

It would seem we have almost everything at our disposal in the modern world, yet even in wealthy countries—where, in theory at least, people have access to all kinds of consumer goods—many still have a sense of emptiness; people feel like they've been robbed. That may be because there's a real price to pay for worshipping things. According to a study by JWT, a New York marketing firm, "Fame and fortune have replaced faith and family as the core of the American Dream." But this dream is a dangerous illusion. Ultimately, money and ownership are status symbols, and symbols are only the mark of happiness, not the essence of it.

You'll often hear it argued that greed is at the bottom of the problems in the world. That is not completely true. It is our *belief* that we must own things—a nice car, a beautiful home, fashionable clothes—to be respected and deemed successful. This "good life," so the story goes, can be attained if we work hard and keep acquiring. That's the rule of the game. And so people regularly go into debt to buy more things. That debt becomes money, which in turn siphons up to the rich. The rich have the ability to find loopholes (from offshore accounts to tax breaks), making them better at the game. And as a consequence, the divide between the rich and poor grows ever more extreme.

We may think that ownership is the solution, but in many ways it's the problem. And while ownership seems natural, that doesn't mean it's good. Evolution has crafted all kinds of other "natural" traits and behaviours that are now maladaptive, or even criminal. In fact, one way to define civilization is as the shared effort to mitigate the danger of evolved responses.

But there is another issue at the root of ownership. And that is, even if our species blindly believes it owns the world, that doesn't mean it's truly ours.

11
REVOLUTION

A new type of thinking is essential if mankind is to survive and move toward higher levels.

—ALBERT EINSTEIN

IN THE SPRING OF 2014, I packed my bags and travelled to the remote Lamu Archipelago off the coast of Kenya to begin research for this book. I chose the location specifically because Lamu is a living relic; it is a place out of time. Donkeys are still the primary mode of transport on the island, and dhows, with their beige triangular sails, drift across the glimmering waters as they have for two thousand years. By immersing myself in another reality, I hoped I could begin to question my own.

One morning toward the end of my stay, the owner of my hotel told me she was making a trip across the water to a nearby island. She did this occasionally, she said, bringing seed pods with her that had fallen from her baobab trees. As we chatted by my hammock, I looked up at the two massive baobabs above us with their hulking silvery trunks. They looked like giant guards protecting us with their shade against the African sun. These trees were still babies, however, perhaps only one or two hundred years old. I asked the owner why she made the effort to plant trees that she would never see mature in her lifetime. She replied that the trees were a gift to people of the future so that they too might sit under giant baobabs one day and simply admire them.

After she left, I returned to my hammock and did just that. Looking up, I thought how few people today would question the right to own something like a tree. If it's on your land, it's yours; it belongs to you. You can let it grow or you can chop it down. Because you own it, you can do what you choose. But that morning, sitting beneath those African giants, I was struck by how strange it was to think one could own something like a baobab tree. How could I own a life that will outlive me by two thousand years? Compared to the baobab, I felt like a mayfly. At that moment, the thought that this tree could be mine seemed utterly absurd.

Back in Toronto, I found myself thinking in a similar way about ownership. All around me, people were said to be living in a "housing bubble." But owning a house is different from owning a tree. Yes, some homes are investments, but our primary homes are our shelters; they are not optional, they are a necessity. As I thought more about it, I realized we have relationships to our homes, and whether we own them or rent them (and most people don't even own their home, the bank does) makes no difference to our *feeling* of ownership. People love their homes. We tend to them, whether it's painting the walls, mowing the grass, or remodelling the kitchen. Some homes, like our childhood homes, have a significant place in our hearts. But does that really require ownership? Do we need to own something for it to be a part of us?

As I look around my home now, the same could be said of the things that I feel are most precious to me, like photographs or heirlooms. These objects are cherished because they serve as an unbroken chain of time, they are intergenerational, and priceless. My grandmother's watch, for instance, can't do any of the things a modern watch can. It can't track my fitness, make phone calls, or receive emails. It's a treasured object not because of what it does, and would be even if it stopped ticking. It keeps time for me in a different sense. It is the physical carrier of a memory. After all, what would it mean if I lost it? Would it be my grandmother's loss or mine? Or the next generation's?

Once we begin to see how we look at the world through a lens of ownership, we can see how it shapes everything about our reality. It is so deeply intertwined into daily life that it feels like the most natural thing in the world, and we don't question it. After all, all of us, myself included, are owners to some degree. But what does it really mean to *own* something? Is ownership a fundamental, intrinsic reality, like an atom? Or is it just a way of looking at things?

WE HAVE EYES, but that doesn't mean we see clearly. In 1951, Solomon Asch conducted a famous experiment to illustrate this point. The study involved a vision test with fifty college students in Swarthmore, Pennsylvania. As the subjects entered a room, they sat amongst what they thought were seven other students, who were really actors.

In this setting, the "students" were all given a task: to match the length of a line on the left side of the board with a line from a series of different-sized lines, marked A, B, and C, on the right. The test subject was unaware that the others taking the test had been instructed in advance to all choose the same line, which was either too long or too short. It was a trick. So, for example, looking at the illustration below, the group would say that the line on the left matched the line marked A. An answer that was obviously not correct.

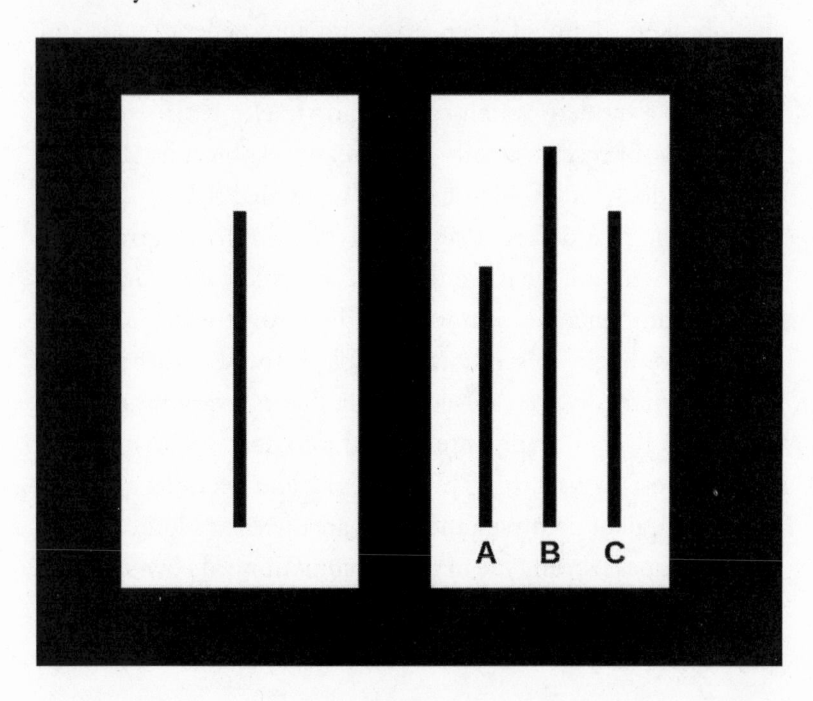

The study found that, faced with wrong answers by their peers, 75 percent of the test subjects conformed to the group's choice at least once; that is, they claimed to see what the others saw, even though the answer was incorrect. For participants in the control group, where no actors were present, fewer than 1 percent gave the wrong answer.

In an effort to find out what was happening in the brain when subjects took this test, researchers at Emory University modified the Asch experiment by placing them inside an fMRI scanner to see which parts of their brains were activated during the task. They were asked to take a similar vision test, but instead of lines they were shown 3-D objects. The researchers expected that if conformity was the result of conscious decision-making, then the prefrontal cortex would be activated, as this is the area of the brain related to planning, decision-making, and moderating social behaviour. What they found was a surprise: conformists showed activity in the parietal and occipital areas of the brain. This is where sensory information and sight are processed, indicating to the team that the conformity was not just a decision, it was having a perceptual influence on sight. Said in another way, conformity may have altered their perception. For these subjects, it wasn't a case of seeing is believing, but rather the reverse.

The Asch experiment on conformity has been much publicized over the years, but one aspect of the study is rarely emphasized: how often people *refused* to conform. While it is true that 75 percent of subjects in the original study went along with the majority at least once, it is also true that 95 percent of the subjects "rebelled" at least once and stood by what they saw. Further, 25 percent of participants refused to be swayed at any time.

Two hundred years earlier, David Hume roughly predicted those results. In fact, our eagerness to conform was one of his key insights into human behaviour. And this conformity is not benign; it has enormous political implications. Writing in his 1741 treatise *Of the First Principles of Government*, Hume observed,

> Force is always on the side of the governed, the governors have nothing to support them but opinion. It is therefore, on opinion only that government is founded, and this maxim extends to the most despotic and most military governments, as well as to the most free and most popular.

That being the case, tyrants depend on public endorsement just as much as elected prime ministers and presidents do. But even where there is a majority there will always be objectors. Indeed, the 25 percent that refuse to conform, that refuse to be blindly ruled, are why we are subject to ever-growing surveillance.

So what of the majority? What causes their conformity? Brain research has shown that there's a price to be paid for independent thinking. In the Asch fMRI study, subjects who refused to conform showed activation in an area of the brain where other subjects did not: the amygdala, the part of the brain associated with "fight or flight" responses. Standing up for your beliefs, then, has a cognitive cost, as standing against a consensus can mean a conflict. For social animals like us, that causes anxiety and distress. Ultimately, standing up against the majority requires a fair amount of bravery.

THE WORD "APOCALYPSE" in ancient Greek sounds slightly less foreboding when you know its etymology. In the original definition, an apocalypse is an "uncovering" of knowledge, a lifting of the veil, a revelation. In essence, it is the dawning of clarity. This kind of revelation is what philosophers, sages, and scientists have long been calling for: for humanity to rub the sleep out of its eyes and begin to see things as they really are, to recognize that what we call reality is in fact an illusion.

Many great thinkers have written about the reality bubble. In Plato's allegory of the cave, prisoners watched shadows projected against the cave wall and came to believe the shadows were the real thing; they confused appearance with reality. The ancient Indian texts of the Upanishads introduced the concept of *maya*, the veil obscuring the true, eternal world. And in Buddhist philosophy, the foundational principle of *dharma*, or cosmic law, leads practitioners on a quest to see reality as it is, rather than as we perceive it to be, and understand that in this bigger picture everything is connected.

To see the world clearly, we must first become aware of the veil; we must recognize our blind spots. The way we've come to perceive reality is so deeply ingrained, so socially and intergenerationally enrooted, that we've lost sight of the manner in which we think. This is important, because what we think creates reality. Clock time, with the five-day workweek and the nine to five of the "real" world, exists not because of some cosmic temporal order but because we invented it, we maintain it, and it's become the reality we adhere to.

Inheriting a reality makes it that much harder to see it for what it is. As Peter Berger and Thomas Luckmann write in *The Social Construction of Reality*, "If one says, 'This is how these things are done,' often enough one believes it oneself. An institutional world, then, is experienced as an objective reality. It has a history that antedates the individual's birth and is not accessible to his biographical recollection. It was there before he was born, and it will be there after his death."

Our constructed world has become so real and dear to us, we've forgotten that what we call reality is a product of our minds. This collective amnesia is perhaps not so surprising considering the decades we spend educating and socializing the young. We expect our youth to grow up and conform like subjects in the Asch experiment, to see a reality that isn't really there. It's ironic, then, that we say children live in a world of make-believe, because in truth, adults do too. The difference is, children can tell you that their world is made up, whereas adults cannot.

This make-believe world is now so powerful that even its antecedent, the natural world, has become its hostage. As Yuval Noah Harari writes in *Sapiens*, we lived in the past in a dual reality: "On the one hand, the objective reality of rivers, trees and lions; and on the other hand, the imagined reality of gods, nations and corporations. As time went by, the imagined reality became ever more powerful, so that today the very survival of rivers, trees and lions depends on the grace of imagined entities such as gods, nations and corporations."

It is with these manufactured entities that we legitimize our prerogative over nature. After all, that's what gods, nations, and corporations do. They give us legitimacy. They back up the belief that *Homo sapiens* own the whole world.

Only one species believes it owns the air, owns the water, and owns the land. We have given ourselves the rights to buy and sell space, and buy and sell time. In fact, that is the foundation of the global economy: that we can own the very dimensions we inhabit. But critically, not only do humans own the planet, we own all the life upon it. Our species alone operates with the belief that *at our discretion* we have the right to buy and sell other species. For us, life itself is a commodity. And with the hyper-accelerated pace of commercial trade, it is unsurprising that life itself is now disappearing.

According to the World Wildlife Fund, by 2020 we will have seen a staggering 67 percent decline in wildlife populations around the planet since 1970.[1] With threats from food systems and agriculture, habitat loss, and species exploitation, more than *half* of vertebrate life—more than half of our wild mammals, birds, and fish—has already gone.

But it's not just the animals. As I type these words, news headlines are revealing a sad fate for the baobab. These ancient behemoths, some standing since the height of the Roman Empire, are dying at an unprecedented rate. Botanist Adrian Patrut believes the most likely culprit is climate change. Since 1960, the number of baobab trees in Africa has halved. According to Patrut, who has been radiocarbon-dating baobabs for over

1 "Populations of vertebrate animals—such as mammals, birds, and fish—have declined by 58% between 1970 and 2012."

fifteen years, it is time to list these long-lived trees as endangered.

That Africa's iconic "tree of life" is dying is symbolic. As my friend the late activist Rob Stewart observed, "By mid-century, if we continue on our current trajectory, we face a world with no fisheries, no coral reefs, no rainforests, declining oxygen concentrations, and nine billion hungry, thirsty people fighting over what remains. . . . In the lifespan of one [baobab] tree, we've consumed most of our life support system."

This stark vision of the future has put fear in the hearts of even our most dystopian science fiction writers. As William Gibson has darkly observed, few people today even consider writing about a future beyond 2100. In an interview in *Vulture* magazine, he said, "What I find far more ominous is how seldom, today, we see the phrase 'the 22nd century.' Almost never."

If we *are* to survive into the twenty-second century, we will need a new global model. We must throw off the constraints of past political ideologies, whether left or right, because they all begin with the wrong question. They ask *who* should have the right to own the world, not *if* we should have that right at all.

YOU OFTEN HEAR PEOPLE say we need to fight the system, or that the system is broken. But what is the "system," exactly?[2] *Where* is the system?

The system, as I've argued in this book, is our life-support system. It is the system we have built so that we no longer have to rely on the whims of nature's cycles. It is the system that made us the most powerful species on earth. And while it would be easy to assume that the goal of our system is our species' survival,

---

2 According to Donella Meadows, a pioneer of systems thinking, a system can be defined as "a set of things—people, cells, molecules, or whatever—that are interconnected in such a way that they produce their own pattern of behavior over time."

it is not. If it were, then every human being would have enough food and energy, and enough time and space, to thrive. But we know that this is not the case. The irony is that our survival is merely incidental to the goal of the system: ownership. The real goal is simply to own as much as possible: to own time, to own space, to own food, to own energy, to own everything except our waste. This is the model that runs the world. A system in which nature's gifts are no longer free. And now, to acquire its goods, we must sell the most precious thing we are born with: our time.

There is however, another critical factor that eludes us and that is where the system is. We cannot see the system because it exists in our blind spots. It is nature in disguise. Today, if we fail to see our connection to the natural world, it's because most of our products look nothing like it: a chicken nugget does not look like a bird; coal does not look like an ancient forest; and fertilizer bears no resemblance to air. Nature has been transformed into a product. In fact, every year it's transformed into trillions and quadrillions of products. These feed our booming population and voracious desires, which leads us to plunder these natural "resources" at an ever increasing rate. As a consequence, the economy grows, but nature dies. And while we are clever animals, none of us could have foreseen the twist in the plot. None of us could have guessed that in the end we will need to pull the plug on our own life-support system, and if we don't, it will destroy us.[3]

---

3 In a letter entitled "World Scientists' Warning To Humanity" more than 1,500 leading scientists and Nobel laureates put their names to the following caution:

> "Human beings and the natural world are on a collision course . . . If not checked, many of our current practices put at serious risk the future that we wish for human society and the plant and animal kingdoms, and may so alter the living world that it will be unable to sustain life in the manner that we know.

The threat is just as real as the system is real. Real in the sense, to paraphrase Philip K. Dick, that our problems are not going to go away if we just stop believing in them. But systems—as solid as they may seem—are still built upon our collective thinking, our ways of seeing the world. That being the case, they *can* change, but only if we change the thinking behind them. As Robert Pirsig put it in *Zen and the Art of Motorcycle Maintenance*, "If a factory is torn down but the rationality which produced it is left standing, then that rationality will simply produce another factory. If a revolution destroys a government, but the systematic patterns of thought that produced that government are left intact, then those patterns will repeat themselves."

What we need, urgently, is a way out of this hall of mirrors. And we can find it with science, because science can shatter old world views. It can literally change the world by changing the way we see.

The greatest minds in history were rebellious thinkers like Galileo, Darwin, and Einstein. We know their names because they were bold, scientific revolutionaries who defied majority opinion and reshaped our understanding of the world. We are the lucky inheritors of their radical thought. Galileo proved that Earth revolves around the sun, and now we know we are not the centre of the universe. Darwin connected the dots of life, proving that animals are our relatives, that we are part of a long evolution of life, not separate from but rather connected to all other living beings. And Einstein flipped the script on the dimensions, proving that space-time is relative to the observer and there is no such thing as fixed, absolute time or space.

These big shifts in thinking could hardly be derived from common sense. In fact, they run counter to what we perceive about the world with our physical senses. Writing about Carl Sagan's work, literary critic Maria Popova likewise observed that,

> We navigate the world by our common-sense perception, but that perception has blinded us to reality again and again. We have mistaken our sensorial intuitions for facts of the universe—for millennia, we held wrong beliefs about Earth's shape, motion, and position, because it feels flat and static beneath our feet, and central to the order of the cosmos. We have mistrusted processes and phenomena beyond the boundaries of what we can touch and feel with our limited senses—from evolution, which unfolds on scales of time too vast to be visible within a human lifetime, to quantum mechanics, which operates on subatomic scales imperceptible and almost inconceivable to the human observer.

Our senses *tell us* we are separate from the universe and the environment and other living beings. Science, however, presents the evidence to *prove* that our physical perceptions are wrong. That is the great gift of science and scientists: they are reality testers; they can pierce through our blind spots with evidence that provides a clearer, more objective view of the world. The greatest scientists in the world are remembered because they burst our reality bubble.

It's been said that hindsight is 20/20, and nowhere is this truer than in science, where old ideas about reality seem ludicrous to us now. In his groundbreaking book *The Structure of Scientific Revolutions*, Harvard-trained physicist Thomas Kuhn says he found his inspiration to write after studying the work of Aristotle. Kuhn observed that the intellectual giant "appeared not only ignorant of mechanics, but a dreadfully bad physical scientist as well." Further, "About motion, in particular, his writings seemed to me full of egregious errors, both of logic and of observation."

Kuhn's key insight was that a brilliant thinker like Aristotle, by modern standards, looked like a rambling idiot. But Aristotle was operating within a given scientific paradigm; his ideas were shaped by a very particular world view. This was Kuhn's epiphany, one that led him to coin the term "paradigm shift." If, in the past, scientific knowledge had been seen as a slow but accumulative process that moved toward a greater understanding of physical reality, Kuhn showed that in fact knowledge grows in giant, discontinuous leaps. Or, to use a different analogy, a caterpillar doesn't grow into a butterfly, it enters a chrysalis stage, where it dissolves into a genetic soup that reforms into a very different-looking insect but one that still has a memory of its earlier being.

Scientific revolutions, for Kuhn, are these dramatic shifts in thinking. But he was keen to point out that scientific advances are not like visual gestalts, for example, where one's perception of a single image can shift back and forth between two seemingly different things. It is a bigger shift, he writes, than mere illusion: "The marks on paper that were first seen as a bird are now seen as an antelope, or vice versa. That parallel can be misleading. Scientists do not see something as something else; instead, they simply see it." The distinction is critically important. As the philosopher of science Ian Hacking notes, "The cautious will gladly say that one's view of the world changes, but the world stays the same. Kuhn wanted to say something more interesting. After a revolution, scientists, in the field that has been changed, work in a different world."

We too live in a different world because of these scientific revolutions. Our collective minds have changed because of what we've learned. Though there are, of course, those who refuse to see what science sees, who trust only their human senses: the flat-earthers and creationists who expect all the fruits of the modern world while refusing to abandon long-obsolete beliefs.

Humanity's greatest thinkers, however, are those who push the boundaries of sight; they are visionaries in the truest sense of the word because they see what for the rest of us is invisible. For Newton, it was the invisible force of gravity; for Van

Leeuwenhoek, it was the invisible animalcules; for Copernicus and Galileo, it was the invisible movement of our still Earth around the sun. As Kuhn was aware, scientists regularly work with "theoretical entities, such as electrons, [observing things] at which one cannot point"; they regularly work in a world that is invisible.

Consequently, there is often a gap between what science sees and the lay person understands. With access to modern technological tools, scanning electron microscopes, mass spectrometers, and fMRIs, scientists can see what the rest of us cannot, and this, in addition to highly focused expertise, results in a significant knowledge gap between scientists and the public. In a recent poll by the Pew Research Center and the American Association for the Advancement of Science (AAAS), the majority of Americans, 79 percent, agreed that scientists and scientific knowledge are invaluable, and yet similar polls have demonstrated that a significant number do not rely on science to support their own views. In a 2013 poll, for instance, only 33 percent of the general public believed climate change to be a serious problem, compared to 77 percent of AAAS scientists, a yawning gap of 44 percentage points.[4] Another reason for the big difference has to do with how science is communicated to the general public. Here, even language can have a significant effect. When scientists say there is "uncertainty" about an effect, the public takes that to mean a lack of knowledge, where a better translation of the term as it is used scientifically would be "range." Similarly, when scientists speak of "positive feedbacks" when it comes to climate change, the public thinks of a good result or positive praise, when in practice it means a self-reinforcing cycle.

---

4 The share of the general public saying that global warming is a very serious problem has fluctuated in Pew Research polling between a low of 32 percent in 2010 to a high of 47 percent in 2009.

Big ideas also take time to transmit, because people are stubborn in their beliefs. Even a century after Copernicus's death, his bold ideas had few converts. And while Newton's groundbreaking proofs were well documented in his *Mathematical Principles of Natural Philosophy*, it was over half a century before his ideas were generally accepted. Nobel Prize–winning physicist Max Planck similarly lamented that "a new scientific truth does not triumph by convincing its opponents and making them see the light, but rather because its opponents eventually die, and a new generation grows up that is familiar with it."

Though Planck was no doubt right, we don't have a generation to figure this out. And Planck didn't have the fortune to live in the high-speed, connected world of today, where we have the power to read, share, and communicate new ideas with one another instantly.

SINCE 1972, ONLY TWENTY-FOUR PEOPLE have gone beyond low Earth orbit and seen the planet whole. NASA's space shuttle missions, along with expeditions to the International Space Station and China's Tiangong stations, have increased that number, but still only just over five hundred people have had the grand privilege of seeing Earth from space. Or put another way: 0.0000072 percent of the human population has seen this magnificent view.

Of those who have made that journey, some have reported a massive shift in their perspective. It even has a name, the overview effect, a "space consciousness" that causes a deep and profound change in thinking allowing the astronauts to see their earthly home in a new way. Physicians who studied the returning travellers reported that "for many of them, it's given them a particular attitude about themselves and their relationship to others. Some became more aware of the earth itself, all of them have developed a new sense of the order of the universe. And those of us who have been close to them have developed the same kind of reactions even though we haven't been there ourselves." Unlike us, on the International Space Station (ISS), men and women witness sixteen sunrises and sunsets a day. But what's even more incredible is that as they look down, our earthly clocks and borders lose their meaning, even though the ISS is a mere 400 kilometres away.

Astronauts also literally see the revolutions of our planet. They see it spin beneath their feet, and their eyes take in the magnitude of its beauty along with the scale of its destruction. In one revolution of Earth, they can bear witness to the deforestation, the droughts, the wildfires, the melting icecaps, the hurricanes, and the pollution. Up in space, the human footprint on Earth is not some abstract thing. It is not data. It is visible.

From space, even the bubble is real. You can see it as the thin bluish-white curve that protects us from the radiation of outer space. Known as the atmosphere, it is the bubble that protects all life on Earth. But the bubble is also a trap. And as scientists tell us, the rate at which CO_2 is increasing in the atmosphere is accelerating, and it is largely human-caused emissions of these heat-trapping gases that are responsible for global warming.

But it doesn't require a space-bound epiphany for us to be aware of what we are doing to our home. And while the overview effect occurs for some, in truth it is a phenomenon not experienced by many astronauts. As the astronaut Chris Hadfield once told me, it is not the cupola view from the International Space Station that changes your perspective but rather your own thoughts and life experience. In other words, you don't need to go to space to see the world differently. You can see it with fresh eyes from right here.

IN TRANSFORMING OURSELVES, we transform the world. In transforming the world, we transform ourselves. As Joseph Campbell detailed in his book *The Hero with a Thousand Faces*, this theme can be found throughout the ages, in many parts of the world. It is the universal story of the hero's journey, and it underlies our most powerful epics, from ancient Greek myths to Hollywood blockbusters like *Star Wars*, *The Lord of the Rings*, or *The Matrix*.

In essence, the hero's journey unfolds in a cycle, or a single revolution. It begins with the hero living an ordinary life in an ordinary world, only to have it all upturned one day when they discover that the world they took for granted has suddenly changed. They find themselves in "an unfamiliar, special world" and must cast aside the status quo, become a seeker of new knowledge. This knowledge is used to battle the trials and challenges the hero will face as their normal way of life increasingly comes under threat. At a certain point, all may seem lost and defeat may seem inevitable, but then, in the final moments, a new insight or revelatory power takes hold, allowing the hero to prevail and make a triumphant return. Now, the hero heads home, this time embracing a new outlook. And while the world may still outwardly appear the same, for the hero it has completely changed.

It's almost as though these epic stories have been preparing us for this very moment. We have reached a time where each of us must rise to the challenges that confront us. It is time for *us* to change. And while most people still live in the "ordinary world," for those who can see, it's clear that impending doom lurks not far away. Indeed, the cracks in our normalcy are already appearing. Scientists tell us we are on the brink of devastating changes and that the world we live in will soon be under siege. Moreover, if we fail to respond, we will be facing not only localized disasters in the coming decades but civilizational catastrophes.

On the other hand, it's something of a cosmic joke that we are all here to face this moment at all. Because frankly, the odds against it are stupendous. As Stephen Hawking pointed out in *A Brief History of Time*, the existence of life on Earth required implausibly perfect conditions in the universe: "If the rate of expansion one second after the Big Bang had been smaller by even one part in a hundred thousand million million, it would have re-collapsed before it reached its present size." In a similar vein, biologist Ken Miller writes, "If g [the gravitational constant] were smaller, the dust from the big bang would just have continued to expand, never coalescing into galaxies, stars, planets, or us. The value of the gravitational constant is just right for the existence of life. A little bigger, and the universe would have collapsed before we could evolve; a little smaller, and the planet upon which we stand would never have formed."

These are just two examples from over two hundred physical parameters in the solar system and universe that needed to be almost perfect for life to have had a chance to evolve. But the odds against your particular existence are even greater.

Ali Binazir, a graduate of Cambridge University, decided to calculate the odds of any one of us being born. Factoring in the chances of your parents meeting (one in twenty thousand) and the chances of them staying together in order to conceive you (one in two thousand), the basic odds of your birth start at around one in forty million. But that's before including biological odds. With your mother producing over one hundred thousand eggs in her lifetime and your father producing four trillion sperm in his, the chances of you, in particular, being here (as opposed to your brother or sister) works out to about one in four hundred quadrillion.

But we have to look back further than just your parents, because genetically you belong to an unbroken chain of familial lineage that goes back over 150,000 human generations. Binazir works out those odds to be in the order of 1 in $10^{45,000}$, too long a number to write out fully on this page, or even this chapter. In fact, it's a number "not just larger than all of the particles in the universe—it is larger than all of the particles in the universe *if each particle were itself a universe.*" To put that in perspective, the probability of your existence is equivalent to "the probability of 2 million people getting together . . . each to play a game of dice with trillion-sided dice. They each roll the dice and they all come up with the exact same number." Meaning, "The odds that you exist at all are: basically zero."

Think about that. In the grand scheme of reality, you have arrived on Earth, at the right place, at the right time, only to appear *exactly* on the eve of the planetary apocalypse?

Really, it's too perfect. Even Hollywood couldn't dream up a better plot. And you, the hero of this story, couldn't find yourself in a more epic, exquisite, or extraordinary tale.

ACKNOWLEDGEMENTS

Writing a book is a solitary process but it is not something you can do alone. First, I am indebted to my brilliant editor Nick Garrison. He, along with my indefatigable agent Rick Broadhead, were the first to believe in this book. Thank you both for your wisdom, excellence, advice and kindness. I would also like to thank the entire Penguin Random House Canada team, specifically: Kristin Cochrane and Nicole Winstanley, Kara Savoy, Tonia Addison, Paisley McNab and Scott Loomer, and my terrific copy editor Alex Schultz. This being my first book, all of their support has been deeply appreciated.

I am also very grateful to the scientists, academics, journalists, researchers and friends who shared their time and expertise. My thanks to: Mark Abbott, Nobu Adilman, Malcolm Clench, Tim Cockerill, Biella Coleman, Martin Fowler, Jonas Frisén, Michael Gillard, David Grimm, Jay Ingram, Peter Jakobs, Naomi Klein, Arthur and Marilouise Kroker, Jo-Anne McArthur, Alan Nazerian, Dan and Shelby Riskin, Bob Rutledge, Joel Solomon, Jan Sorgenfrei, Nigel J.T. Smith, David Suzuki, Astra Taylor and WWF Canada.

For a decade, I had the good fortune of working at Discovery Canada. My thanks go out to all of my CTV/Discovery colleagues, but especially Seonaid Eggett, Kelly McKeown, John

Morrison, Agatha Rachpaul, and Ken Shaw for his encouragement.

To the Black Sheep who inspire me, thank you. And to my closest friends (you know who you are—and why for privacy reasons I won't name you), I love you. McLean Greaves and Rob Stewart, you are both in my heart, always.

Finally, this book is for my family. For their patience, for their never-ending support, for their love, and for being the greatest source of my happiness. I owe you everything—and yet nothing, because you've taught me that the most precious gifts in life are free.

A FULL LIST OF REFERENCES can be viewed online. Please visit http://www.penguinrandomhouse.ca/realitybubble

INDEX